Earth's Crust and Its Evolution - From Pangea to the Present Continents

Edited by Mualla Cengiz
and Savaş Karabulut

Published in London, United Kingdom

IntechOpen

Supporting open minds since 2005

Earth's Crust and Its Evolution - From Pangea to the Present Continents
http://dx.doi.org/10.5772/intechopen.95644
Edited by Mualla Cengiz and Savaş Karabulut

Contributors
Cyril C. Okpoli, Emilio Herrero-Bervera, Michael A. Oladunjoye, Adamu Abubakar, Othniel K. Likkason, Talat Kangarli, Tahir Mammadli, Fuad Aliyev, Rafig Tofig Safarov, Sabina Kazimova, Edmore Utete, Yasuto Itoh, Lawrence Hutchings, Shou-Chneg Wang, Steve Jarpe, Sin-Yu Syu, Kai Chen, Chao-Shing Lee, Saidi Abdollah, Khan Nazer Nasser, Hadi Pourjamali Zahra, Farzad Kiana, Suresh Kannaujiya, Tandrila Sarkar, Abhishek Kumar Yadav, Paresh N.S. Roy, Charan Chaganti, Vikas Adlakha, Kalachand Sain

Notice
Statements and opinions expressed in the chapters are these of the individual contributors and not necessarily those of the editors or publisher. No responsibility is accepted for the accuracy of information contained in the published chapters. The publisher assumes no responsibility for any damage or injury to persons or property arising out of the use of any materials, instructions, methods or ideas contained in the book.

First published in London, United Kingdom, 2022 by IntechOpen
IntechOpen is the global imprint of INTECHOPEN LIMITED, registered in England and Wales, registration number: 11086078, 5 Princes Gate Court, London, SW7 2QJ, United Kingdom
Printed in Croatia

British Library Cataloguing-in-Publication Data
A catalogue record for this book is available from the British Library

Additional hard and PDF copies can be obtained from orders@intechopen.com

Earth's Crust and Its Evolution - From Pangea to the Present Continents
Edited by Mualla Cengiz and Savaş Karabulut
p. cm.
Print ISBN 978-1-83969-077-8
Online ISBN 978-1-83969-078-5
eBook (PDF) ISBN 978-1-83969-079-2

We are IntechOpen,
the world's leading publisher of
Open Access books
Built by scientists, for scientists

6,000+
Open access books available

148,000+
International authors and editors

185M+
Downloads

Our authors are among the

156
Countries delivered to

Top 1%
most cited scientists

12.2%
Contributors from top 500 universities

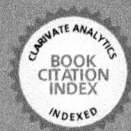

Interested in publishing with us?
Contact book.department@intechopen.com

Numbers displayed above are based on latest data collected.
For more information visit www.intechopen.com

Meet the editors

Dr. Mualla Cengiz is a professor in the Department of Geophysical Engineering, Istanbul University–Cerrahpaşa. She researches paleomagnetism applied to the tectonic assemblages of Anatolia and its surrounding area. She lectures on gravimetry and magnetism of the Earth, rocks, and the environment. She has authored many international publications and collaborated on several national and international projects. She is actively researching geophysical applications in caves, salt resources, and seawater intrusion.

Dr. Savas Karabulut was an assistant professor in the Department of Seismology, İstanbul University during 2002–2016. He has been a lecturer in the Department of Civil Engineering, Gebze Technical University, Turkey since 2021. His research interests are engineering seismology, soil dynamics, seawater intrusion, paleomagnetism, soil engineering problems, statistical hazard analysis, and cave/sinkhole investigation using geophysical methods. He has collaborated on many projects about earthquake mitigation, seismic hazards, urban transformation, and earthquake disaster. He has authored many scientific publications and books and more than 100 conference papers.

Contents

Preface

The supercontinent Pangaea was formed by continental blocks that broke up about 250 million years ago to form Gondwana and Laurasia. The idea that Earth's continental blocks once formed a single supercontinent was proposed in 1912 by the German meteorologist Alfred Wegener in his "continental drift" theory in his 1915 book, *The Origin of the Continents and Oceans*. Wegener believed that this continental drift explained why the borders of South America and Africa fit together. He also highlighted the similar rock formations and fossils on these two continents. His idea, however, lacked a suitable mechanism explaining how the continents move. The geodynamic processes of the opening and closing of oceans along old orogenic belts were described by Wilson (1966) in his article "Did the Atlantic Close and then Re-Open?" published in *Nature*. Wilson argued that the Appalachian–Caledonide belt along the eastern margin of North America and Western Europe was formed by a "Proto-Atlantic Ocean." The idea that radioactive minerals within the mantle produce radiogenic heat as the driving force of convection currents was proposed by Arthur Holmes. In 1960, Harry Hess proposed his theory of seafloor spreading to account for the origin of oceanic ridges. In 1963, the Vine–Matthews–Morley hypothesis confirmed Hess' hypothesis. From 1950 to the 19070s, new data obtained from magnetic and paleomagnetic studies gave numerical parameters on the orientation of continental blocks.

The development of earth science over the last century has made it possible to image the Earth's interior using geophysical and geological data and computer modeling of magnetic, electric, and gravitational fields, as well as the propagation of seismic waves and the formation and deformation of rocks. Plate displacement across plate boundaries was successfully determined with GPS networks, while the development of computer systems gave rise to accurate plate reconstructions. A powerful geophysical imaging method is seismic tomography which enabled the mapping of lateral heterogeneities due to seismic velocity.

This book presents research methods based on field and laboratory studies combined with geodynamic and tectonic outputs. Section 1, "Crustal Evolution and Tectonic Problems," includes three chapters. Chapter 1 examines the tectonic evolution of the Himalayas since the Paleoproterozoic Era. Chapter 2 discusses the geodynamic evolution of the Iranian block during the Cretaceous Period and the importance of the Sabzevar-Nain Basin according to the results of a geological field study. Chapter 3 presents a case study on applying the phase stripping method to the tectonic phases of Southwest Japan to characterize the active fault structures. Section 2, "Geophysical Methods in Geological Applications," includes three chapters. Chapter 4 is a study showing a compilation of heat flow measurements, seismic tomography, seismicity, and hot spring distribution in Taiwan for understanding the relationship between the tectonic and geothermal potential in the study area. Chapter 5 presents an extensive paleomagnetic study carried out on Precambrian basement rocks in southwestern Nigeria to show the paleomagnetic pole movement. Chapter 6 shows the structural features of the Sokoto Basin in northwestern Nigeria using a potential field gravity method. Section 3, "Seismic Forecasting, Seismotectonics and Geodynamic Evolution of the Himalayan Belt," includes three chapters. Chapter 7 presents a seismic hazard assessment done using GPS data in the Northwest and Central Himalayan regions. Chapter 8 applies a Brownian passage-time distribution as a seismic forecasting model for earthquakes. Finally, Chapter 9 discusses seismic risk in eastern Caucasia.

Dr. Mualla Cengiz
Professor,
Istanbul University-Cerrahpaşa,
Department of Geophysical Engineering,
Istanbul, Turkey

Dr. Savaş Karabulut
Profesor,
Gebze Technical University,
Department of Civil Engineering,
Kocaeli, Turkey

Crustal Evolution and Tectonic Problems

Chapter 1

Crustal Evolution of the Himalaya since Paleoproterozoic

Vikas Adlakha and Kalachand Sain

Abstract

Understanding the crustal evolution of any orogen is essential in delineating the nomenclature of litho units, stratigraphic growth, tectonic evolution, and, most importantly, deciphering the paleogeography of the Earth. In this context, the Himalayas, one of the youngest continent-continent collisional orogen on the Earth, has played a key role in understanding the past supercontinent cycles, mountain building activities, and tectonic-climate interactions. This chapter presents the journey of Himalayan rocks through Columbian, Rodinia, and Gondwana super-continent cycles to the present, as its litho units consist of the record of magmatism and sedimentation since ~2.0 Ga. The making of the Himalayan orogen started with the rifting of India from the Gondwanaland and its subsequent movement toward the Eurasian Plate, which led to the closure of the Neo-Tethyan ocean in the Late-Cretaceous. India collided with Eurasia between ~59 Ma and ~40 Ma. Later, the crustal thickening and shortening led to the metamorphism of the Himalayan crust and the development of the north-dipping south verging fold-and-thrust belt. The main phase of Himalayan uplift took place during the Late-Oligocene-Miocene. This chapter also provides insights into the prevailing kinematic models that govern the deep-seated exhumation of Himalayan rocks to the surface through the inter-play of tectonics and climate.

Keywords: Himalaya, crustal evolution, supercontinent cycles, tectonics

1. Introduction

"Present is the key to past," the fundamental *"Uniformitarian Principle"* given by Scottish geologist James Hutton [1] holds even today after two centuries when we try to understand the Earth's crust and its evolution in various orogenic belts. Nature has preserved the information about the history of the Earth in various rocks that geologists extract through advanced techniques of geochemistry, geochronology, thermochronology, structural geology, stratigraphy, geophysics, glaciology, climatology, and atmospheric sciences. In this context, Himalayan orogen plays a significant role in understanding the crustal evolution as its rocks provide a vast range of magmatism and sedimentation records from ~2000 to 8 Ma [2–5]. Thus, Himalayan rocks are not only significant in the understanding of past supercontinental cycles but also play an important role to evaluate feedback processes between the lithospheric deformation, atmospheric circulation, tectonic uplift, global climate change, exhumation, and erosion from millennial to decadal scales [6–12]. Such feedback lays the foundations to understand and mitigate natural hazards, such as floods, landslides, and earthquakes, which bear societal relevance. Given

this background, the present chapter focuses on the crustal evolution, deformation, and exhumation of the Himalayan rocks through time since Proterozoic. We evaluate and summarize how the Himalayan rocks have evolved since the Columbian supercontinental cycle to the loftiest and tallest mountain belt of the world in Cenozoic based on geochronological, structural, metamorphic, and thermochronological record. In addition, we also provide insight into the formation of the India-Asia collision zone that resulted from the continent-continent collision between the Indian and Eurasian/Asian Plate through the closure of the Neo-Tethyan Ocean in the Late-Cretaceous-Eocene.

2. Overview of the Himalayan orogen

The Himalayan orogen is one of the youngest continent-continent collisional mountain belts on the Earth [13]. The orogen is a part of the greater Himalayan-Alpine system, which extends from the Mediterranean Sea in the west to the Sumatra arc of Indonesia in the east over a distance of >7000 km. The composite belt had evolved since the Paleozoic when the Tethyan Ocean closed between two converging landmasses of Eurasia and India. The collision of India and Eurasia took place between ~59 Ma and <40 Ma [14–23] and it was brought about by rifting of India from Africa and East Antarctica during the Mesozoic. The convergence of the Indian landmass is continuing toward the north relative to stable Eurasian landmass forming an orogenic wedge to the south of the Tibetan Plateau [24].

The Himalayan mountain belt with ~2500 km long arc stretches between the structural syntaxial bends of Nanga Parbat in the west to the Namche Barwa in the east (**Figure 1**). The Gangdese Shan, Karakorum Mountains, also known as Trans-Himalaya [25], and Tibetan Plateau lie in the north of the Himalaya. In the west of the range lies the Hindu Kush Mountains and to the east, the Indo-Burma ranges, also known as the Rongklang range [9, 26]. The Indo-Gangetic plains/depression lies in the south of the raised Himalayan front. The arc can be further divided into western, central, and eastern sectors (**Figure 2**). The width of the Himalayan Mountain

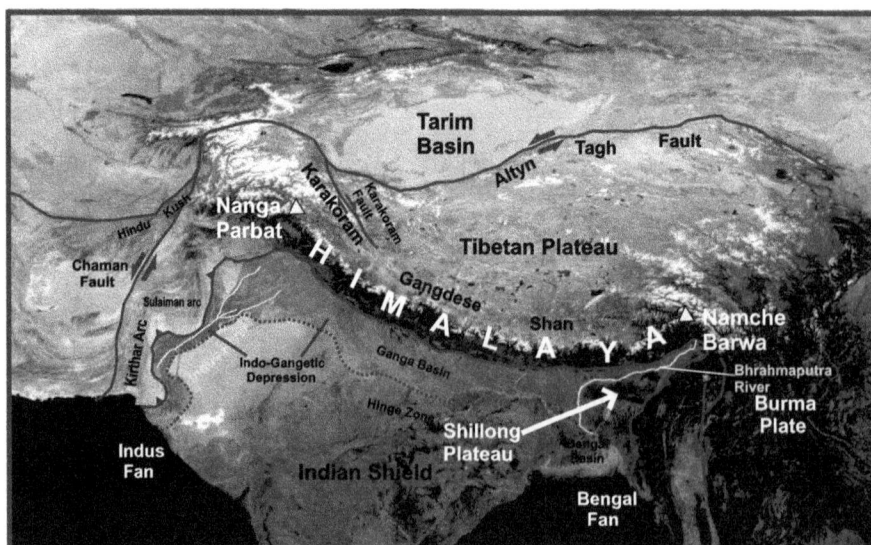

Figure 1.
Topographic map of Himalayan orogen (source: www.wikimapia.org).

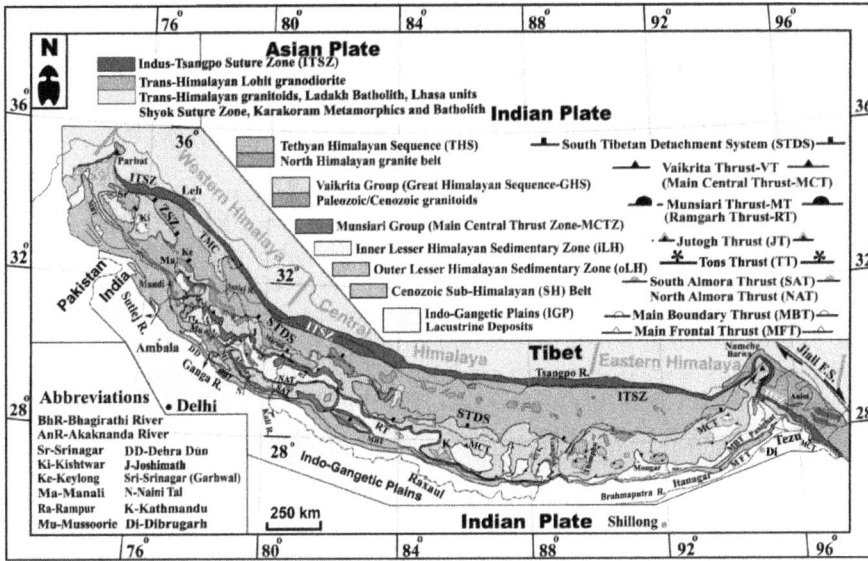

Figure 2.
Geological map of the Himalaya showing main tectono-stratigraphic units (after ref. [3]).

is at its narrowest (100–150 km) in the central sector [9]. The longitudinal river system in the central Himalayas is responsible for the presence of the world's deepest canyons and eight out of the 10 highest mountain peaks on Earth. However, the average relief is significantly less in the western and eastern Himalayas.

3. Geological setting

The Himalaya has been divided into four tectonostratigraphic domains along the entire arc between Indo-Gangetic Plains in the south and Indus Tsangpo Suture Zone (ITSZ) in the north [3, 10, 27–31]. These are (a) Sub-Himalaya, (b) Lesser Himalayan Sequence (LHS), (c) Greater Himalayan Sequence (GHS)/Higher Himalayan Crystallines (HHC), and (d) Tethyan Sedimentary Zone (TSZ) (**Figure 2**). These domains are characterized as south-vergent fold and thrust belts that have emerged as a result of crustal shortening and thickening.

3.1 Sub-Himalaya

The Sub-Himalaya represents the southernmost part of the Himalayan orogen. The Main Frontal Thrust (MFT) separates it from the Indo-Gangetic Plains to the south, while the Main Boundary Thrust (MBT) separates it from the LHS to the north (**Figure 2**). It comprises marine sediments of Paleocene-Eocene and sediments from the continental origin of Miocene-Pliocene [32]. The marine sediments comprising shale, sandstone, and limestone are known as Subathu Formation, while the sediments of continental origin are characterized as Dagshai, Kasauli, and Siwalik group of rocks. The sedimentation record of Subathu formation is described to be ~61.5–43.7 Ma from magnetostratigraphic data [32]. The age of Dagshai formation has been estimated to be ~32–25 Ma, followed by overlying Kasauli formation of ~32–22 Ma and Siwalik sediments of ~14 Ma [33–37]. The thickness of this sequence is ~9–10 km, which was deposited by southward flowing river systems of the Himalaya, forming the erosional history of the orogen [38]. The MFT, bounding the Himalayas to the south is commonly expressed as a zone of folds and blind thrusts [39, 40], which was active during the Pliocene-Holocene [41].

3.2 Lesser Himalayan sequence

The LHS is bounded by MBT to the south and Main Central Thrust (MCT) zone to the north and consists of three sub-units from south to north [3, 42]. These are: (a) Outer Lesser Himalayan (oLH) belt; (b) Lesser Himalayan Crystalline (LHC) nappe, and (c) Inner Lesser Himalayan (iLH) belt (**Figure 2**). The oLH represents Neoproterozoic-Paleozoic-Mesozoic-Eocene sedimentary sequence between MBT and the Tons Thrust (TT)/North Almora Thrust (NAT). These sequences are locally known as Shimla-Jaunsar (comprising of Mandhali, Chandpur, and Nagthat Formations)-Blaini-Krol-Tal Groups in the NW Himalaya [43]. The detrital zircon U-Pb geochronological ages with the oLH belts are 0.95 Ga in the oldest Mandhali Formation, 0.88 Ga in Chandpur Formation, 0.82 Ga in Nagthat Formation, 0.70 Ga in Balini diamictite and Krol sandstone, and 0.525 Ga in lower Tal trilobite bearing strata [44–47]. These ages provide the maximum timing of their deposition from the source rock and are considered synchronous with the Paleozoic magmatism in the source region.

The LHC nappes are basically the synformal klippe that are thrust over LHS and are equivalent to the GHS rocks, which form their root zone. These are locally named as Jutogh-Ramgarh-Almora-Askot-Chiplakot nappe in the western Himalaya and Kathmandu nappe in the central Nepalese Himalaya. The mylonite orthogneiss of Kulu-Ramgarh Nappe provides the oldest zircon U-Pb age of ~1.85 Ga, which is overlain by the Nathuakhan Formation of ~0.80 Ga [48, 49]. The Almora nappe in the Kumaun region of western Himalaya is characterized by equivalent ~1.85 Ga of mylonite granite-gneiss at its base and younger populations of ~0.85 to 0.58 and 0.55 Ga from garnetiferous-quartzite-schists and intrusive granites, respectively [49, 50].

The iLH belt is the Paleoproterozoic meta-sedimentary sequence between TT/NAT and MCT and represents the oldest and lowermost sequence of the LHS that was deposited between ~1.90 and 1.80 Ga as constrained by detrital zircon Geochronology [42, 47, 48, 51]. It is noteworthy that the rocks of iLH are characterized by fewer older minor peaks also between ~2.4 and 2.6 Ga [2, 52–59]. These rocks are locally known as Rautgara-Gangolihat-Deoban-Berinag Groups in the western Himalaya, Kushma Group in Central Nepal, and Daling-Shuma Groups in the Eastern Himalayas of Bhutan and Arunachal Pradesh [43, 53, 55, 60]. It is significant to note that there has been a stratigraphic break of nearly ~1 Ga between the timing of deposition of the iLH and oLH [3].

3.3 Greater Himalayan sequence/higher Himalayan Crystallines

This sequence represents the Himalayan orogen's backbone, which exhibits the most uplifted and most eroded part of the orogen. The sequence is bounded by the South Tibetan Detachment System (STDS) in the north, which separates it from the TSZ. The Munsiari Thrust (MT)/MCT forms the southern boundary of this sequence, where it abuts against the iLH rocks. The ~15 to 20 km thick sequence is divided into two main groups: (a) Munsiari Group/MCT zone and (b) Vaikrita Group. The Munsiari Group overrides the iLH along the MT/MCT 1 or lower MCT (in Nepal Himalaya)/MCT [3, 9, 23, 25, 43, 61–63] and contains mylonitized and imbricated Paleoproterozoic megacryst granite gneiss, fine-grained biotite paragneiss, garnetiferous mica schist, phyllonite and sheared Amphibolite [3]. Based on the geochemistry and geochronological studies, these rocks have been part of the Paleoproterozoic magmatic arc [42], as most of the zircon U-Pb ages of these rocks lie between ~1.97 and 1.75 Ga [55, 59, 60, 64–69], that is, similar to iLH rocks.

The Vaikrita Group consists of amphibolite facies to migmatitic ortho- and para-gneisses rocks and characterizes typical inverted metamorphism [70–80]. The Vaikrita Thrust (VT)/MCT 2 or upper MCT (in Nepal Himalaya) forms the base of this group while these rocks abut against the TSZ rocks along the STDS. This group is different from the Munsiari Group and iLH rocks as the rocks of the Vaikrita Group provide characteristic Neoproterozoic zircon U-Pb ages between ~1.05 and 0.80 Ga with fewer peaks at ~2.50 and ~ 1.80 Ga [3, 4, 60, 69, 81–83].

The GHS generally forms a continuous belt along the entire length of the orogen. Still, it also occurs as isolated patches surrounded by low-grade Tethyan strata, such as in the Zanskar and Tso Morari regions of NW India and in the Nanga Parbat massif of northern Pakistan [84–86]. This coeval slip along the MCT and STDS during ~20 to 15 Ma is responsible for the ductile extrusion of the GHS/HHC rocks between these bounding fault zones [30, 87–89]. It is noteworthy that all the north-dipping faults in the Himalaya sole into a mid-crustal décollement at depth, the Main Himalayan Thrust [MHT, 90], which lie over the Indian basement.

Apart from the Munsiari and Vaikrita Group of rocks, the presence of Cambro-Ordovician granitoids are unique within the HHC (**Figure 2**). These granitoids belong to Pan-African magmatism and lie to the north of *sensu-stricto* the MCT [9, 91]. These are locally named Central gneiss, Dalhousie, Chauri, Dhauladhar, Palampur, Mandi, Pandoh, and Karsog Granite [27, 92–94]. However, these Paleozoic granites occur sparsely also within the Lesser Himalaya, Tethys Himalaya in the Karakoram and Tibet [95–97]. Geochemical analysis of these granitoids suggests that these rocks were formed in a syn-collision environment and have peraluminous (S-type) and mildly metaluminous (I-type) affinities [98]. Few occurrences of these granitoid bodies exhibit mild alkaline nature that was formed in a post-collision, anorogenic setting [99, 100]. In general, these granitoids belong to early Paleozoic magmatism (ca. 475 Ma) as reported through whole-rock Rb-Sr isochron age U-Pb zircon geochronological data [4, 91, 101–103].

The Proterozoic rocks of GHS have undergone crustal thickening and short-ening, metamorphism, and partial melting during Himalayan orogeny, that is, post-India-Asia collision [9]. The leucocratic magmatism in the Himalayas, mainly by muscovite dehydration melting, can be traced along the entire arc of the orogen in the GHS and as well as TSZ [60, 69, 104–106]. In the late stage, the high-grade metamorphic rocks of the GHS exhumed to the surface through the interplay of tectonics and climatic processes [10–12, 107].

3.4 Tethyan sedimentary zone (TSZ)

Late Precambrian to Eocene siliciclastic and carbonate sedimentary rocks interbedded with Paleozoic and Mesozoic volcanic rocks are exposed to the north of GHS, mainly forming the TSZ [9, 108–119]. It is bounded in the north by Great Counter Thrust where it is juxtaposed with Tso Morari Crystallines and/or ITSZ rocks, and in the south by north-dipping STDS. The STDS is locally named as Zanskar Shear Zone (ZSZ) [79, 120], Rohtang Shear Zone (RSZ) in Himachal Pradesh [121], Trans-Himadri Fault (THF) in Kumaon Himalaya [122], and STDS in Nepal Himalaya [87, 123]. The TSZ has been divided into four subsequences [9]: (a) Early Cambrian to Devonian pre-rift sequence characterized by lithologic units deposited in epicratonal setting, (b) Carboniferous-Lower Jurassic rift and post-rift sequence, (c) Jurassic-Cretaceous passive continental margin sequence, and (d) Cretaceous-Eocene syn-collisional sequence. These rocks form the cover sequence of the GHS and are also known as Haimanta Formation in the NW Himalaya that yields detrital zircon U-Pb ages between 0.55 and 3.0 Ga [3, 69, 121].

4. Crustal evolution of Himalayan rocks since Paleoproterozoic

4.1 Paleoproterozoic

The iLH and MCT zone represent the oldest terrane of the Himalayan rocks that were formed during the Paleoproterozoic. The rocks of both of these terranes have been hypothesized to be a part of an Andean-type arc system that formed during the assembly of the Columbian supercontinent at ~1.9 Ga (**Figure 3a**) [3, 42, 124–128]. The assembly of Columbia involved North America (NA), Eastern Antarctica (EA), North China (NC), and India (I) continents that formed the arc system. The granitoids of the MCT Zone/Munsiari Group were formed due to hydrous partial

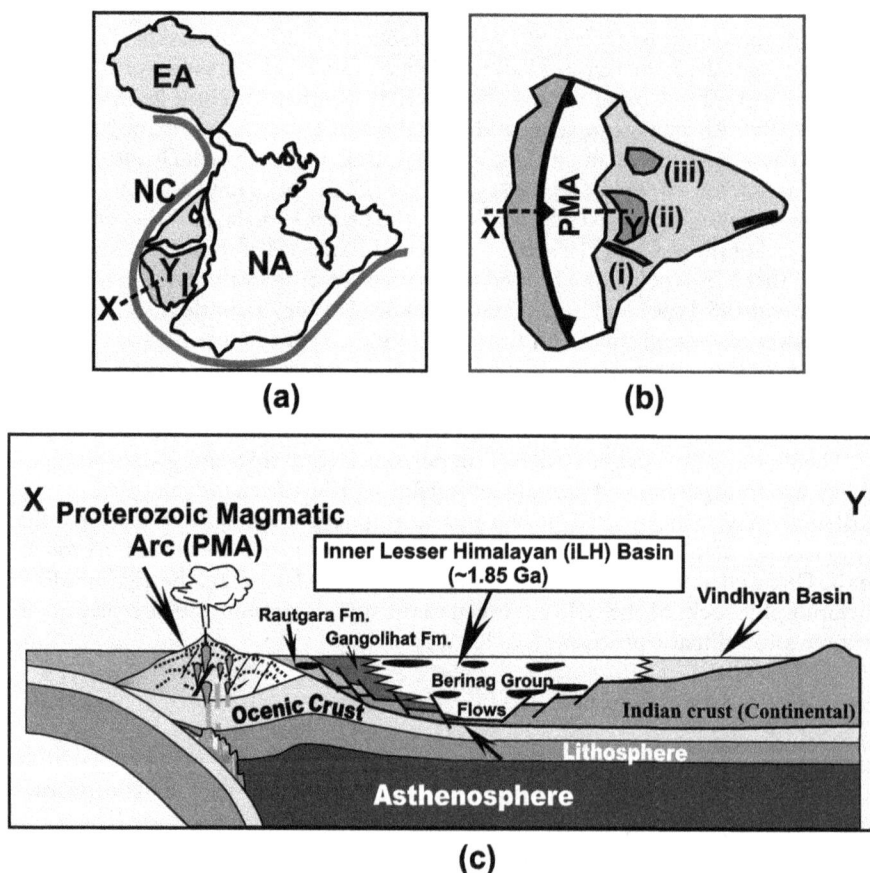

Figure 3.
Reconstruction of Columbia supercontinent ca. 1.9 Ga showing position of India and the Proterozoic magmatic arc (after ref. [3]). (a) Reconstruction showing continents of NA-North America, EA-eastern Antarctica, NC-North China, I-India. (b) Position of the Proterozoic magmatic arc, its configuration of forearc (blue), backarc (yellow), and the position of India with the (i) Aravalli, (ii) Bundelkhand, and (iii) Singhbhum cratons. Cross-section along XY. (c) Reconstruction showing the evolution of the MCTZ and iLH as the Paleoproterozoic magmatic arc and backarc basin, respectively, during ~2.0–1.8 Ga. Subducted and partially melted oceanic lithosphere caused the emplacement of arc granitoids. The iLH sediments were deposited in rifted back-arc basin and received sediments, from both the arc, that is, MCT zone and Indian craton.

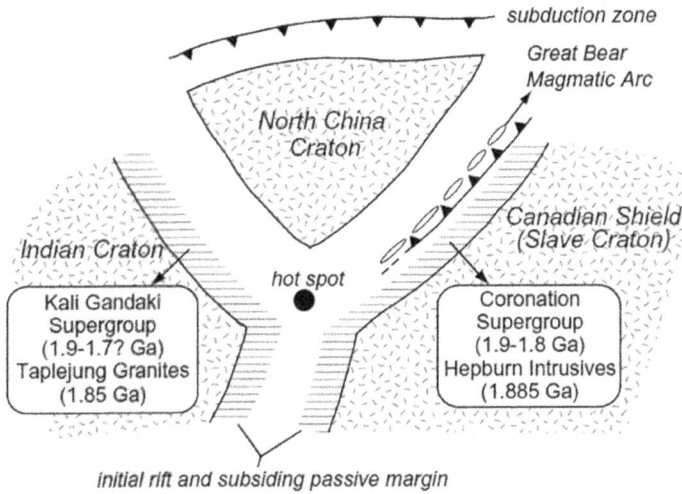

Figure 4.
A possible model for the deposition of iLH and MCT zone rocks in a rift and passive continental margin set up as originated from the mantle plume (after ref. [129]).

melting of the older crust that involved mafic sources and sediments from the subduction zone (**Figure 3b** and **c**). The volcano-sedimentary sequence of iLH, forming Rautgara-Gangolihat-Berinag, their equivalent formations, was deposited during ~2.0–1.8 Ga in the rifted back-arc basin (**Figure 3c**) [3]. The sediments in this back-arc basin were supplied from both the Proterozoic magmatic and Indian Shield (**Figure 3c**). However, some workers believe that the iLH and MCT zone rocks were part of a rift and passive continental margin set up that was originated from the mantle plume (**Figure 4**) [129, 130]. In this hypothesis, the ~2.0 to 1.8 Ga rocks of the iLH and MCT zone are considered equivalent to the Paleoproterozoic Coronation Supergroup in the Wopmay orogen, northwest Canada [131, 132].

4.2 Neoproterozoic

The Columbian supercontinent broke up during the Neoproterozoic, resulting in the separation of the Indian craton from Columbia, which thus later became a passive margin along its northern limit. India had reassembled again within a short duration of 1 Ga at ca. 1.1–0.9 Ga with Madagascar, Seychelles, Karakoram Terrane/Pamir, Tarim in the west, South China in the north, Australia in the east, and East Antarctica in the southeast, respectively, forming the Rodinia Supercontinent (**Figure 5**). The Neoproterozoic granitoids have been recognized in the GHS throughout the Himalayan arc and are associated with the Rodinia Supercontinent assembly [60, 66, 81, 134–137]. Apart from the magmatic origin of granitoids within the GHS, the sediments containing detrital zircons of ca. ~1.1 to 0.8 Ga in the Vaikrita Group and the oLH sequences have been sourced from (a) Within-basin magmatic bodies, that is, granites intrusive and orthogneisses, for example, Peshawar, Black mountain of Western Himalaya syntaxis, Chor region of Himachal Himalaya, and Cona, Bhutan and Hapoli regions of Arunachal Himalaya [3], (b) "In-board" Indian Craton, that is, Aravalli Delhi Mobile Belt (ADMB) and Central Indian Tectonic Zone (CITZ), which is collectively known as Great Indian Proterozoic Fold Belt (GIPFOB, **Figure 6**) [3], and (c) external "Outboard" terranes of Nubian-Arabia, Africa, Madagascar, eastern Antarctica, Australia, that is, those belonged to Rodinia Supercontinent assembly [45, 53, 81], through erosion and transportation of sediments by long paleo-river systems.

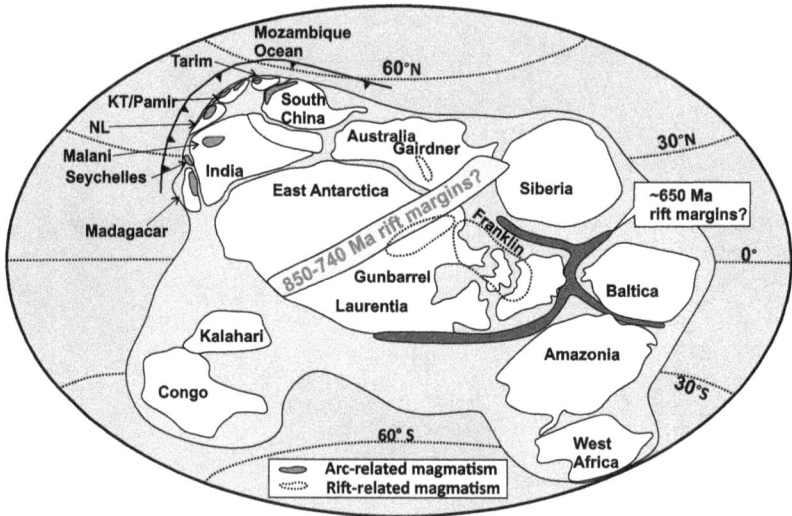

Figure 5.
Reconstruction of Rodinia supercontinent showing the position of India (after refs. [4, 133]).

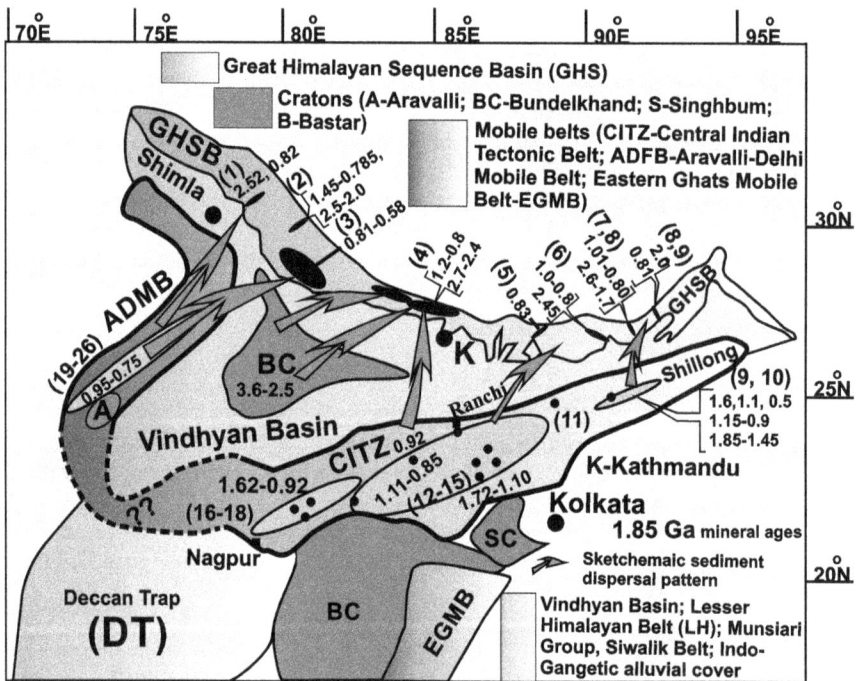

Figure 6.
Neoproterozoic detrital zircon in the great Himalayan sequence (GHS) and correlatable successions in the lesser Himalaya are sourced from various parts of the Indian craton (after ref. [3]).

4.3 Late Neoproterozoic to Cambro-Ordovician

Rodinia supercontinent broke up during 750–600 Ma, which led to the pathway for the formation of Gondwanaland during the Cambrian–Ordovician (**Figure 7**) [e.g., 133, 138, 139]. India was a part of the Gondwanaland that also consisted of South

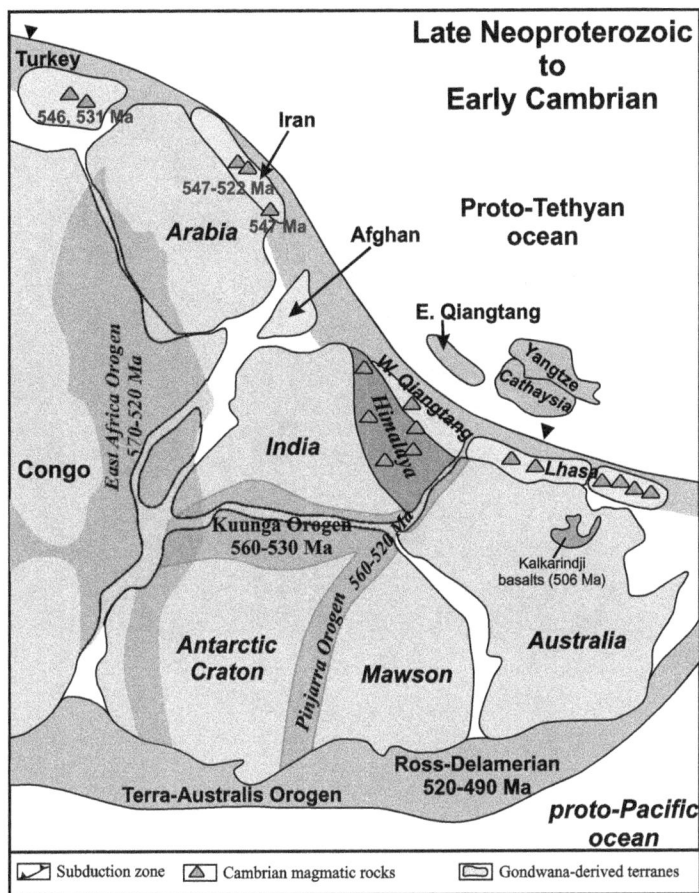

Figure 7.
Position of India during the Gondwana supercontinent assembly in Cambro-Ordovician (after refs. [4, 91]).

America, Africa, Madagascar, Australia, and Antarctica [140]. Together, these continents formed a subduction system along the northern margin of the Gondwanaland [141–144]. Thus, a thermal event associated with the Pan-African orogeny during the Cambro-Ordovician resulted in the formation of granitoids, such as Dalhousie, Chauri, Dhauladhar, Palampur, Mandi, Pandoh, and Karsog Granite, presently within the GHS and TSZ of the Himalayan arc. Many researchers have proposed the Cambrian-Ordovician event as the pre-Himalayan metamorphic event that resulted in the crustal anataxis of the local Neoproterozoic crustal rocks during syn-to post-collisional crustal thickening, leading to the generation of S-type granitoids [91, 145–153].

4.4 Silurian to cretaceous

The Gondwanaland was the southern part of the most recent supercontinent Pangea. The Pangea attained its condition of maximum packing at ~250 Ma and started breaking up during ~250 to 230 Ma (**Figure 8**) [140, 154, 155]. The northern part was named Laurasia or Mega Laurasia and contained the northern continents—North America, Greenland, Europe, and northern Asia. The present-day Karakoram Terrane forms the south-western margin of the Tibetan Plateau. It is equivalent to the SE-Pamir terrane and Central Pamir terrane in the west and

Figure 8.
Paleogeographic map showing the break-off of the Cimmerian terranes from Pangea (based on refs. [140, 154, 155]).

Qiangtang terrane in the east, which belonged to Gondwanan ancestry (**Figure 9**). These terranes of Central Pamir, SE-Pamir, Karakoram, and Quiangtang got separated from Gondwana during Permian due to the rifting process that formed the part of the Cimmerian belt [156]. This event resulted in the opening of the Neo-Tethys Ocean (**Figure 10a**). This event was followed by the accretion of these terranes of the Cimmerian belt with the Asian Plate along the Jinsha Suture Zone (JSZ) during the Middle-Cretaceous or maybe earlier [157, 158], resulting in the closure of Paleo-Tethys Ocean (**Figures 9** and **10b**). Initially, the Central Pamir and SE-Pamir were accreted along the Rushan-Pshart suture during Triassic-Jurrasic (**Figure 9**), with slightly later accretion of Southern Pamir and the Karakoram along Tirich Boundary Zone (TBZ) (**Figure 9**) [159].

The Neo-Tethyan ocean closed due to the rifting of the Indian Plate from Gondwana and its subsequent journey toward the Asian Plate (**Figure 10b**). The interoceanic Dras volcanic island arc was formed by the initial subduction within the Neo-Tethyan Ocean during Middle Jurassic (**Figure 11a**) [23]. The closure of the Neo-Tethyan ocean resulted in the subduction of the Neo-Tethyan oceanic lithosphere below the southern margin of the Asian Plate along the Shyok Suture Zone (SSZ) in the north-western domain of Karakoram and along the BNS in central Tibet (**Figure 11b** and **c**) [160, 161]. The formation of calc-alkaline continental arc magmatism at ca. ~205–100 Ma due to the subduction of Neo-Tethys oceanic lithosphere along the SSZ produced Karakoram Batholith on the southern margin of the Asian Plate [160–165]. The final collision between the Indian and Asian plates occurred along the ITSZ that was accompanied by the formation of the Kohistan-Ladakh arc (KLA). Subsequently, the formation of ophiolitic and sedimentary sequences took place along the ITSZ (**Figure 11**) [23, 162, 166, 167]. The KLA

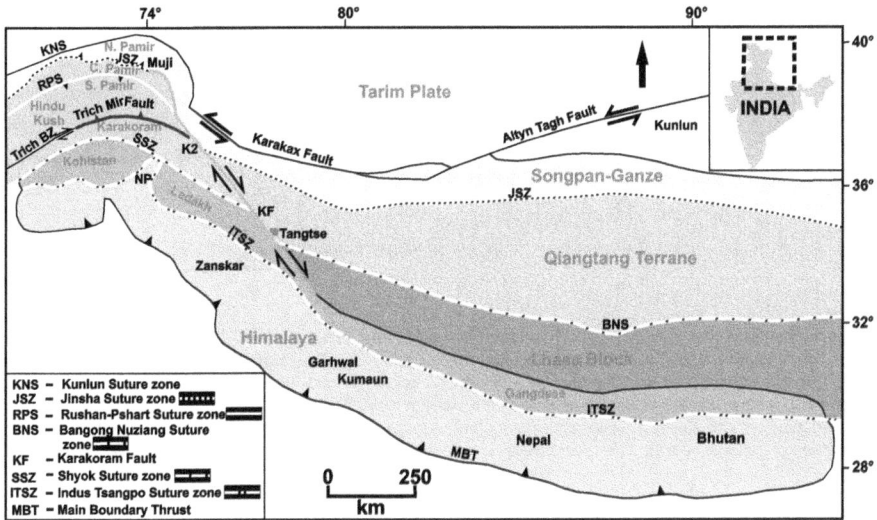

Figure 9.
Tectonic map of the Himalayan-Tibetan orogenic belt showing a present-day configuration of the Karakoram, SE-Pamir terrane, and central Pamir, which were part of the Cimmerian terrane, rifted from Gondwana and led to the opening of neo-Tethys Ocean (after, ref. [4]).

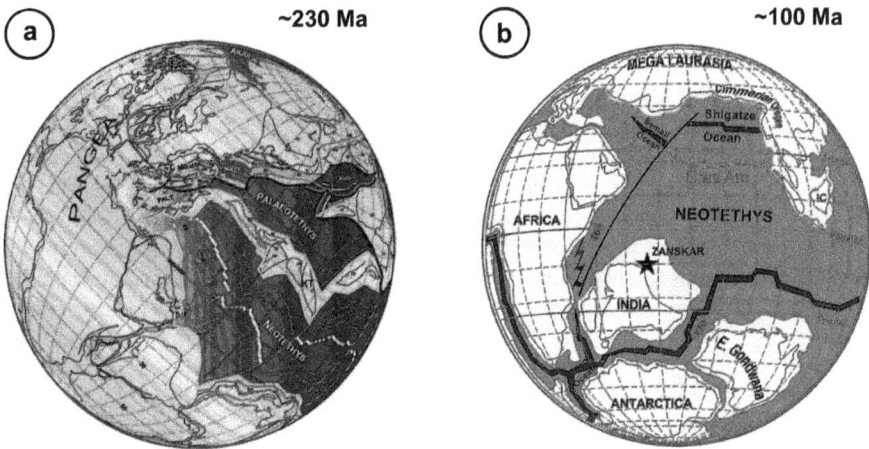

Figure 10.
(a) Palaeotethys Ocean is getting closed as Cimmerian terrane is approaching toward Eurasia. Please note the position of India, that is, the position of rifting from Gondwana land, (b) Cimmerian terrane that includes the Karakoram as a part of Eurasia. Note the formation of Dras island arc. India is moving toward Eurasia (based on refs. [140, 154, 155]).

witnessed a major episode of subduction-related magmatism at ~85 to 40 Ma with small pulses at ~110 to 100 Ma. The magmatism led to the emplacement of Andean-type plutons during the Late Cretaceous to middle Paleogene [23]. These magmatic rocks are collectively known as the Ladakh Batholith in the western Himalayas, Gangdese Batholith in Tibet, and Lohit Batholith in the Eastern Himalayas [9]. The SSZ closed at ca. 85 Ma through the juxtaposition of the Ladakh arc and Karakoram batholiths [160, 168]. The palaeomagnetic anomalies in the Indian ocean suggest that the convergence of the Indian plate slowed down at ~55 ± 1 Ma [18]. The palaeolatitude evidence suggests that Tethyan succession in the Himalayas overlaps with the Lhasa terrane overlap at 22.8 ± 4.20 N palaeolatitude at 46 ± 8 Ma [169, 170].

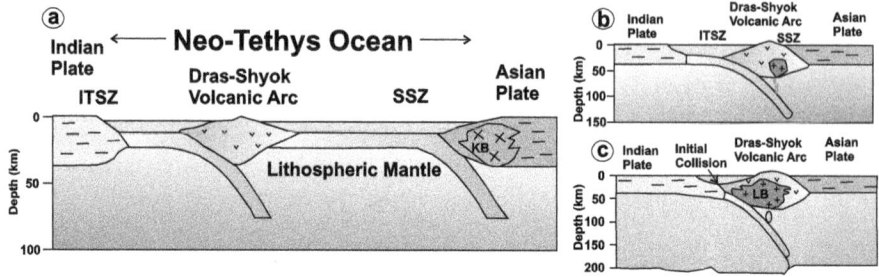

Figure 11.
Schematic model showing the stages for the collision of Indian and Asian plate (a) Dras-Shyok volcanic formed due to subduction of oceanic lithosphere within the neo-Tethys Ocean arc during middle Jurassic to late cretaceous and formation of Karakoram batholith during early cretaceous due to subduction of Tethyan oceanic lithosphere beneath the southern Asian plate margin along SSZ; (b) closure of SSZ due to collision of Dras volcanic arc and Karakoram terrane during late cretaceous; (c) formation of Ladakh batholith due to subduction of Tethyan oceanic lithosphere below Dras volcanic arc during late cretaceous (after ref. [23]).

The final closure of the Indian Plate with the Asian Plate took place along the ITSZ between ~59 Ma to <40 Ma [14–23].

4.5 Cenozoic

The Cenozoic era represents the main phase of Himalayan orogeny. In the initial stage (~45–23 Ma, **Figure 12a**), after the India-Asia collision, crustal shortening and thickening led to the early prograde regional metamorphism mainly during ~45 to 35 Ma under ~8 to 11 kbar and ~ 600 to 700°C [9, 30]. In this phase, thrusting along the STDS started [9, 171]. This was the time when the GHS/LHS were covered by the TSZ and Gondwana sediments. The initiation of MCT led to the emplacement of the GHS over the LHS during the Early Miocene, that is, at ~23 to 18 Ma (**Figure 12b**) [9, 172–174]. The younger event of metamorphism at ~23 to 15 Ma at ~6 to 8 kbar and ~ 500 to 750°C is considered to be synchronous with the ductile deformation along the MCT zone [9, 30]. The intense ductile shearing caused the formation of the inverted metamorphism sequence across the GHS along with Miocene leucogranites generation [9, 78, 104, 175–181]. The MCT forms the roof of the major thrust duplex within the LHS (**Figure 12c**) [60, 83, 174]. The LHS duplex formation led to erosion of the GHS and the formation of the antiformal stack in the form of LH window zones and synformal nappes in the Himalaya, mainly during ~18 to 14 (**Figure 12c**). Thus, the HHC and LH Window zones became the fastest exhuming bodies, with exhumation rates up to ~3 mm/yr., in the Himalayas since Miocene [10–12, 107, 121, 182]. Later, the activation of the MBT at ~10–12 Ma led to the thrusting of the LHS over the Sub-Himalaya [183–185], while the deposition of Siwalik sediments initiated at ~14 Ma (**Figure 12d**). In the Plio-Quaternary, the initiation of MFT took place that forms the southernmost boundary of the Himalaya and juxtaposes the older rocks of Sub-Himalaya along the modern Indo-Gangetic alluvium (**Figure 12e**). In this phase, the MCT also got reactivated in an out-of-sequence manner that led to the rapid exhumation of the rocks of the MCT zone [186–188].

Apart from the aforementioned general model for the Himalayan evolution, it is noteworthy that the ductile extrusion of the GHS has been explained by three main models. The Channel flow-focused denudation model [8] considers the GHS as a partially molten lower/middle crust that extruded southward from Tibet during Eocene-Oligocene via the formation of the pressure gradient between Tibet and India due to the high elevation of the Tibetan plateau (**Figure 13a**). Wedge-extrusion model states that the MCT and STDS form the tapered core (**Figure 13b**). The gravitational

(a) Eocene to Early Miocene (~45-23 Ma)
Thrusting along the STDS and Metamorphism of the HHC

(b) Early Miocene (~23-18 Ma)
Extrusion of the HHC between MCT and the STDS and
onset of duplex structure in the LHMS

© **Middle Miocene (~18-14 Ma)**
Development of Duplex structures within the LHS and folding of the emplaced HHC and MCT.

(d) Middle Miocene to Present (~14 Ma-2.6 Ma)
Reactivation of the MCT as out-of-sequence thrusting in the southern limb of the nappe and
as back-thrusting in the northern limb of the nappe, Onset of thrusting along the MBT

(e) Pliocene-Quarternary (~2.6 Ma-present)
Activation of MFT

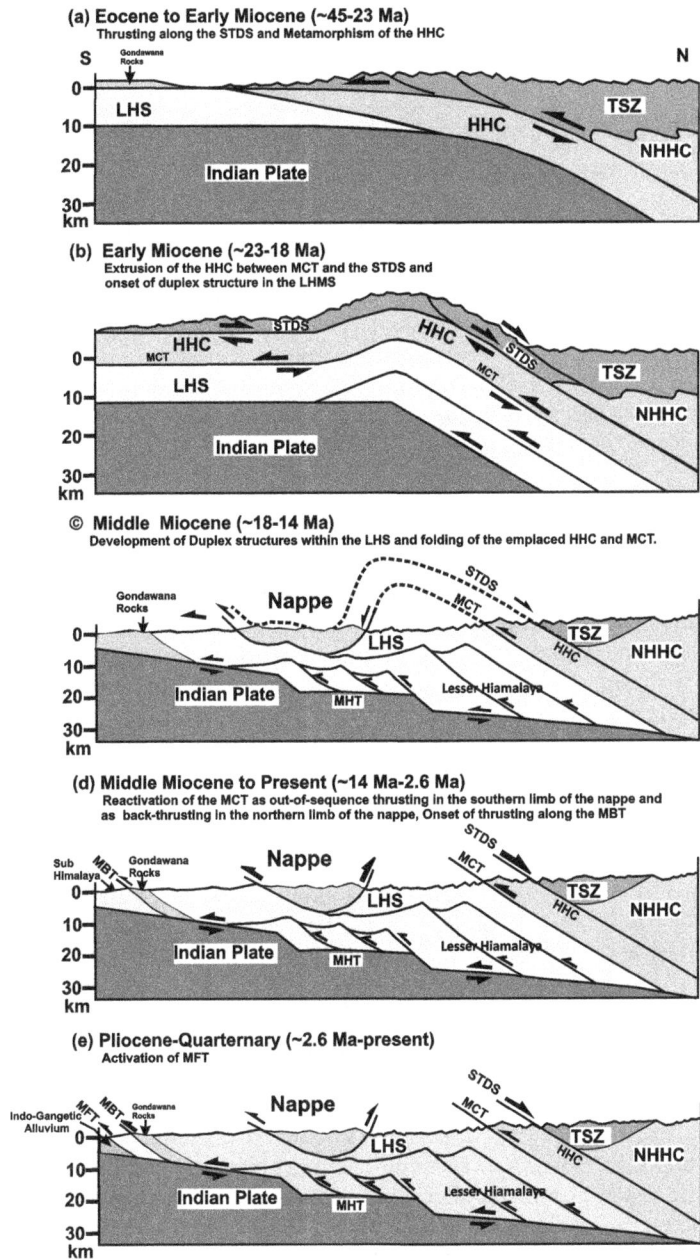

Figure 12.
Sketch showing the general model for the structural evolution of the Himalayan orogen during the Cenozoic (based on refs. [9, 171]).

collapse of over thickened continental crust resulted in the development of the STDS [189]. The tectonic wedging model, in which TSZ abut against the LHS (e.g., in the Himachal Pradesh, India) along the MCT due to its termination against the STDS. Thus, the GHS core remains at depth and subsequently forces itself toward the surface (**Figure 13c**) [190]. Thus, fault kinematics, that is, thrusting, folding, gravitational unloading, the geometry of the subsurface in combination with intense orographic precipitation, controlled the Cenozoic development of the Himalayan orogen.

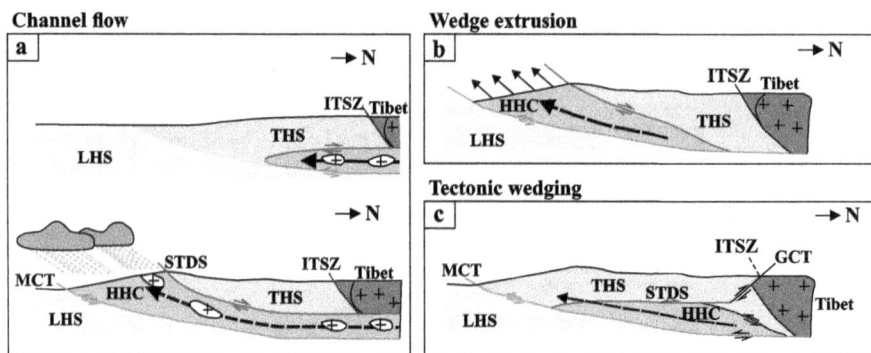

Figure 13.
Tectonic models for the emplacement of the HHC: (a) channel flow model; (b) wedge extrusion model and (c) tectonic wedging (after ref. [69]).

5. Conclusion

The Himalayan crust has evolved through multiple stages of supercontinent cycles since ~2.0 Ga. The iLHS and the MCT zone represent the oldest crust that was formed during the Columbian supercontinent assembly. The rocks of the MCT zone have been derived as Andean-type magmatic arc, while the iLHS was evolved as a back-arc basin, the sediments of which were supplied from both the MCT zone and Indian craton. There is no tectonostratigraphic evolutionary record available for ~1 Ga between the timing of deposition of iLHS rocks (~1.85 Ga) and the deposition of the Vaikrita Group of GHS and oLH (~0.85 Ga). Thus, the northern boundary of India was a passive margin before the deposition of GHS and oLH during the Rodinia supercontinent assembly. The Paleozoic granitoids (~0.48 Ga) within the GHS/TSZ represent the record of pre-Himalayan metamorphism during Pan-African orogeny that formed the Gondwanaland as the southern part of Pangea. The Gondwanaland broke up at ca. 230 Ma as the Cimmerian belt consisting of Central Pamir, SE-Pamir, Karakoram/Quiangtang terranes rifted and moved toward the Asian Plate and led to the closure of Paleo-Tethys ocean and opening of Neo-Tethys ocean. India rifted apart from the Gondwanaland at ~230 to 200 Ma and traveled toward Asia, leading to Neo-Tethys ocean's closure. The Neo-Tethys ocean closed along the SSZ at ~85 Ma by forming the junction between the Asian margin and Dras Island Arc/Ladakh Batholith. The northern margin of the Indian continental crust closed along the ITSZ at <40 Ma, the Ladakh Batholith being its northern boundary. The major event of metamorphism and deformation of the Himalayan crust occurred since Eocene-Oligocene, leading to the formation of north-dipping thrust sheets along with MCT, MBT, and MFT. The fault kinematics, that is, thrusting and folding combined with climatic erosion, led to the exhumation of high-grade metamorphic rocks to the surface [9, 12].

Acknowledgements

We thank the editor Sara Debeuc for the invitation to submit the chapter on Himalayan Crustal Evolution. This work is supported by the CAP Himalaya grant (Activity 7) to V. Adlakha. Prof. A.K. Jain and Shailendra Pundir are thanked for fruitful discussions, informal review, and sharing their figure drafts. Kunal Mukherjee is thanked for extending his help during the final compilation and formatting work. Prof. Nand Lal and R. C. Patel are thanked for their constant encouragement. K. Sain acknowledges the SERB-DST for awarding him with the J.C. Bose National Fellowship.

Conflict of interest

The authors declare no conflict of interest.

Author details

Vikas Adlakha and Kalachand Sain*
Wadia Institute of Himalayan Geology, Dehradun, India

*Address all correspondence to: kalachandsain7@gmail.com

IntechOpen

References

[1] Hutton J. Theory of the Earth. California, USA: Create Space Independent Publishing; 1785

[2] Gehrels G, Kapp P, DeCelles P, Pullen A, Blakey R, Weislogel A, et al. Detrital zircon geochronology of pre-Tertiary strata in the Tibetan-Himalayan orogen. Tectonics. 2011;**30**(5):TC5016

[3] Jain AK, Mukherjee PK, Singhal S. Terrane characterization in the Himalaya since Paleoproterozoic. Episodes. 2020;**43**(1):346-357

[4] Kumar S, Pundir S. Tectono-magmatic evolution of granitoids in the Himalaya and trans-Himalaya. Himalayan Geology. 2021;**42**(2):213-246

[5] Tu JY, Ji JQ, Gong JF, Yan QR, Han BF. Zircon U–Pb dating constraints on the crustal melting event around 8 Ma in the eastern Himalayan syntaxis. International Geology Review. 2016;**58**(1):58-70

[6] Molnar P, England P, Martinod J. Mantle dynamics, uplift of the Tibetan plateau, and the Indian monsoon. Reviews of Geophysics. 1993;**31**(4): 357-396. DOI: 10.1029/93RG02030

[7] Ruddiman WF. Tectonic Uplift and Climate Change. New York: Plenum Press; 1997. p. 535

[8] Beaumont C, Jamieson RA, Nguyen MH, Lee B. Himalayan tectonics explained by extrusion of a low-viscosity crustal channel coupled to focused surface denudation. Nature. 2001;**414**(6865): 738-742. DOI: 10.1038/414738a

[9] Yin A. Cenozoic tectonic evolution of the Himalayan orogen as constrained by along-strike variation of structural geometry, exhumation history, and foreland sedimentation. Earth-Science Reviews. 2006;**76**(1-2):1-31

[10] Thiede RC, Ehlers TA. Large spatial and temporal variations in Himalayan denudation. Earth and Planetary Science Letters. 2013;**371**:278-293

[11] Adlakha V, Patel RC, Lal N. Exhumation and its mechanisms: A review of exhumation studies in the Himalaya. Journal of the Geological Society of India. 2013;**81**(4):481-502

[12] Adlakha V, Sain K, Mukherjee K. Exhumation processes and mechanisms in the Himalayan-Tibetan orogen: A review. Himalayan Geology. 2022;**43**(1B):241-252

[13] Dewey JF, Bird JM. Mountain belts and the new global tectonics. Journal of geophysical Research. 1970;**75**(14): 2625-2647

[14] Aitchison JC, Ali JR, Davis AM. When and where did India and Asia collide? Journal of Geophysical Research: Solid Earth. 2007 May;**112**(B5):B05423. DOI: 10.1029/2006jb004706

[15] Bouilhol P, Jagoutz O, Hanchar JM, Dudas FO. Dating the India–Eurasia collision through arc magmatic records. Earth and Planetary Science Letters. 2013;**366**:163-175. DOI: 10.1016/j. epsl.2013.01.023

[16] Colleps CL, McKenzie NR, Horton BK, Webb AA, Ng YW, Singh BP. Sediment provenance of pre-and post-collisional cretaceous–Paleogene strata from the frontal Himalaya of Northwest India. Earth and Planetary Science Letters. 2020;**534**:116079. DOI: 10.1016/j.epsl.2020.116079

[17] Hu X, Garzanti E, Wang J, Huang W, An W, Webb A. The timing of India-Asia collision onset–facts, theories, controversies. Earth-Science Reviews. 2016;**160**:264-299. DOI: 10.1016/j.earscirev.2016.07.014

[18] Klootwijk CT, Gee JS, Peirce JW, Smith GM, McFadden PL. An early

India-Asia contact: Paleomagnetic constraints from Ninetyeast ridge, ODP leg 121. Geology. 1992;**20**(5):395-398. DOI: 10.1130/0091-7613(1992) 020<0395:AEIACP>2.3.CO;2

[19] Najman Y, Appel E, Boudagher-Fadel M, Bown P, Carter A, Garzanti E, et al. Timing of India-Asia collision: Geological, biostratigraphic, and palaeomagnetic constraints. Journal of Geophysical Research: Solid Earth. 2010;**115**(B12):B12416. DOI: 10.1029/2010JB007673

[20] Rowley DB. Age of initiation of collision between India and Asia: A review of stratigraphic data. Earth and Planetary Science Letters. 1996;**145**(1-4):1-3. DOI: 10.1016/S0012-821X(96)00201-4

[21] Searle MP, Windley BF, Coward MP, Cooper DJ, Rex AJ, Rex D, et al. The closing of Tethys and the tectonics of the Himalaya. Geological Society of America Bulletin. 1987;**98**(6):678-701. DOI: 10.1130/0016-7606(1987) 98<678:TCOTAT>2.0.CO;2

[22] Sen K, Adlakha V, Singhal S, Chaudhury R. Migmatization and intrusion of "S-type" granites in the trans-H imalayan L adakh magmatic arc of north I ndia and their bearing on I ndo-E urasian collisional tectonics. Geological Journal. 2018;**53**(4): 1543-1556. DOI: 10.1002/gj.2973

[23] Jain AK. When did India–Asia collide and make the Himalaya? Current Science. 2014;**25**:254-266

[24] Molnar P, Tapponnier P. Cenozoic tectonics of Asia: effects of a continental collision: features of recent continental tectonics in Asia can be interpreted as results of the India-Eurasia collision. Dcience. 1975;**189**(4201):419-426

[25] Heim A, Gansser A. Central Himalayan Geological observations of the Swiss expedition 1936. Mémoires de

la Société des Sciences Naturelles de Neuchâtel. 1939;**73**(l):1-245

[26] Guzman-Speziale M. Seismicity and active tectonics of the western Sunda arc. The tectonic evolution of Asia. 1996:63-84

[27] Gansser A. Geology of the Himalayas. New York: Wiley InterScience; 1964

[28] Thakur VC, Rawat BS. Geological Map of Western Himalaya (explanation). Dehra Doon: Wadia Institute of Himalayan Geology; 1992. p. 22

[29] Thakur VC. Geology of the Western Himalaya. Vol. 363. New York: Pergamon Press; 1992

[30] Hodges KV. Tectonics of the Himalaya and southern Tibet from two perspectives. Geological Society of America Bulletin. 2000;**112**(3):324-350. DOI: 10.1130/0016-7606(2000) 112<324:tothas>2.0.CO;2

[31] Valdiya KS. The making of India: Geodynamic Evolution. 2nd ed. Berlin: Springer International Publishing AG; 2015. p. 924

[32] Sangode SJ, Kumar R, Ghosh SK. Magnetic polarity stratigraphy of the Siwalik sequence of Haripur area (HP), NW Himalaya. Journal of Geological Society of India (Online archive from Vol 1 to Vol 78). 1996;**47**(6):683-704

[33] Burbank DW, Beck RA, Mulder T. The Tectonic Evolution of Asia. The Himalayan Foreland Basin. Washington, D.C.: AGU Publications; 1996

[34] Najman YM, Pringle MS, Johnson MR, Robertson AH, Wijbrans JR. Laser 40Ar/39Ar dating of single detrital muscovite grains from early foreland-basin sedimentary deposits in India: Implications for early Himalayan evolution. Geology. 1997;**25**(6):535-538

[35] Najman Y, Bickle M, Chapman H. Early Himalayan exhumation: Isotopic

constraints from the Indian foreland basin. Terra Nova. 2000;**12**:28-34

[36] Jain AK, Manickavavasagam RM, Singh S. Himalayan Collision Tectonics. Hauppauge, New York.: Field Science Publishers; 2002

[37] Jain AK, Lal N, Sulemani B, Awasthi AK, Singh S, Kumar R, et al. Detrital-zircon fission-track ages from the lower Cenozoic sediments, NW Himalayan foreland basin: Clues for exhumation and denudation of the Himalaya during the India-Asia collision. Geological Society of America Bulletin. 2009;**121**(3-4):519-535

[38] Srikantia SV, Bhargava ON. Geology of Himachal Pradesh. Bangalore: Memoir Geological Society of India; 1988. p. 406

[39] Nakata T. Active faults of the Himalaya of India and Nepal. Geological Society of America Special Paper. 1989;**232**(1):243-264

[40] Yeats RS, Lillie RJ. Contemporary tectonics of the Himalayan frontal fault system: Folds, blind thrusts and the 1905 Kangra earthquake. Journal of Structural Geology. 1991;**13**(2):215-225

[41] Molnar P. Structure and tectonics of the Himalaya: Constraints and implications of geophysical data. Annual Review of Earth and Planetary Sciences. 1984;**12**(1):489-516

[42] Mukherjee PK, Jain AK, Singhal S, Singha NB, Singh S, Kumud K, et al. U-Pb zircon ages and Sm-Nd isotopic characteristics of the lesser and great Himalayan sequences, Uttarakhand Himalaya, and their regional tectonic implications. Gondwana Research. 2019;**75**:282-297

[43] Valdiya KS. Geology of Kumaun Lesser Himalaya. Dehradun, Uttarakhand: Wadia Institute of Himalayan Geology; 1980

[44] Myrow PM, Hughes NC, Paulsen TS, Williams IS, Parcha SK, Thompson KR, et al. Integrated tectonostratigraphic analysis of the Himalaya and implications for its tectonic reconstruction. Earth and Planetary Science Letters. 2003;**212**(3-4):433-441

[45] Myrow PM, Hughes NC, Goodge JW, Fanning CM, Williams IS, Peng S, et al. Extraordinary transport and mixing of sediment across Himalayan Central Gondwana during the Cambrian–Ordovician. Bulletin. 2010;**122**(9-10):1660-1670

[46] Hofmann M, Linnemann U, Rai V, Becker S, Gärtner A, Sagawe A. The India and South China cratons at the margin of Rodinia—Synchronous Neoproterozoic magmatism revealed by LA-ICP-MS zircon analyses. Lithos. 2011;**123**(1-4):176-187

[47] McKenzie NR, Hughes NC, Myrow PM, Xiao S, Sharma M. Correlation of Precambrian–Cambrian sedimentary successions across northern India and the utility of isotopic signatures of Himalayan lithotectonic zones. Earth and Planetary Science Letters. 2011;**312**(3-4):471-483

[48] Célérier J, Harrison TM, Webb AA, Yin A. The Kumaun and Garwhal lesser Himalaya, India: Part 1. Structure and stratigraphy. Geological Society of America Bulletin. 2009;**121**(9-10): 1262-1280

[49] Mandal S, Robinson DM, Khanal S, Das O. Redefining the tectonostratigraphic and structural architecture of the Almora klippe and the Ramgarh–Munsiari thrust sheet in NW India. Geological Society, London, Special Publications. 2015;**412**(1):247-269

[50] Trivedi JR, Gopalan K, Valdiya KS. Rb-Sr ages of granitic rocks within the Lesser Himalayan nappes, Kumaun, India. Journal of Geological Society of

India (Online archive from Vol 1 to Vol 78). 1984 Oct 1;**25**(10):641-654

[51] Mandal S, Robinson DM, Kohn MJ, Khanal S, Das O, Bose S. Zircon U-Pb ages and Hf isotopes of the Askot klippe, Kumaun, Northwest India: Implications for Paleoproterozoic tectonics, basin evolution and associated metallogeny of the northern Indian cratonic margin. Tectonics. 2016;**35**(4): 965-982. DOI: 10.1002/2015TC004064

[52] Parrish RR, Hodges V. Isotopic constraints on the age and provenance of the lesser and greater Himalayan sequences, Nepalese Himalaya. Geological Society of America Bulletin. 1996;**108**(7):904-911. DOI: 10.1130/ 0016-7606(1996)108<0904:ICOTAA> 2.3.CO;2

[53] DeCelles PG, Gehrels GE, Quade J, LaReau B, Spurlin M. Tectonic implications of U-Pb zircon ages of the Himalayan orogenic belt in Nepal. Science. 2000;**288**(5465):497-499

[54] DeCelles PG, Robinson DM, Quade J, Ojha TP, Garzione CN, Copeland P, et al. Stratigraphy, structure, and tectonic evolution of the Himalayan fold-thrust belt in western Nepal. Tectonics. 2001;**20**(4):487-509

[55] Kohn MJ, Paul SK, Corrie SL. The lower lesser Himalayan sequence: A Paleoproterozoic arc on the northern margin of the Indian plate. Bulletin. 2010;**122**(3-4):323-335

[56] Martin AJ, DeCelles PG, Gehrels GE, Patchett PJ, Isachsen C. Isotopic and structural constraints on the location of the Main central thrust in the Annapurna range, Central Nepal Himalaya. Geological Society of America Bulletin. 2005;**117**(7-8): 926-944

[57] Martin AJ, Burgy KD, Kaufman AJ, Gehrels GE. Stratigraphic and tectonic implications of field and isotopic

constraints on depositional ages of Proterozoic lesser Himalayan rocks in Central Nepal. Precambrian Research. 2011;**185**(1-2):1-7

[58] Martin AJ. A review of Himalayan stratigraphy, magmatism, and structure. Gondwana Research. 2017;**49**:42-80

[59] Khanal S, Robinson DM, Mandal S, Simkhada P. Structural, geochronological and geochemical evidence for two distinct thrust sheets in the 'Main central thrust zone', the Main central thrust and Ramgarh–Munsiari thrust: Implications for upper crustal shortening in Central Nepal. Geological Society, London, Special Publications. 2015;**412**(1):221-245. DOI: 10.1144/ SP412.2

[60] Yin A, Dubey CS, Webb AA, Kelty TK, Grove M, Gehrels GE, et al. Geologic correlation of the Himalayan orogen and Indian craton: Part 1. Structural geology, U-Pb zircon geochronology, and tectonic evolution of the Shillong plateau and its neighboring regions in NE India. Bulletin. 2010;**122**(3-4):336-359. DOI: 10.1130/B26460.1

[61] Shreshtha M, Jain AK, Singh S. Shear sense analysis of the higher Himalayan Crystalline Belt and tectonics of the south Tibetan detachment system, Alaknanda–Dhauli ganga valleys. Uttarakhand Himalaya. Current Science. 2015;**25**:1107-1118

[62] Hashimoto H, Ohta Y, Akiba C. Geology of Nepal Himalaya. Sapporo Japan: Hokkaido University; 1973. p. 281

[63] Arita K. Origin of the inverted metamorphism of the lower Himalayas, Central Nepal. Tectonophysics. 1983;**95**(1-2):43-60

[64] Pathak M, Kumar S. Petrology, geochemistry and zircon U–Pb–Lu–Hf isotopes of Paleoproterozoic granite gneiss from Bomdila in the western

Arunachal Himalaya, NE India. Geological Society, London, Special Publications. 2019;**481**(1):341-377. DOI: 10.1144/SP481-2017-169

[65] McQuarrie N, Long SP, Tobgay T, Nesbit JN, Gehrels G, Ducea MN. Documenting basin scale, geometry and provenance through detrital geochemical data: Lessons from the Neoproterozoic to Ordovician lesser, greater, and Tethyan Himalayan strata of Bhutan. Gondwana Research. 2013;**23**(4):1491-1510

[66] Mottram CM, Argles TW, Harris NB, Parrish RR, Horstwood MS, Warren CJ, et al. Tectonic interleaving along the Main central thrust, Sikkim Himalaya. Journal of the Geological Society. 2014;**171**(2):255-268. DOI: 10.1144/jgs2013-064

[67] Singh SA. A review of U–Pb ages from Himalayan Collisonal Belt. Journal of Himalayan Geology. 2005;**26**(1):61-76

[68] Singh S, Jain AK, Barley ME. SHRIMP U-Pb c. 1860 Ma anorogenic magmatic signatures from the NW Himalaya: Implications for Palaeoproterozoic assembly of the Columbia supercontinent. Geological Society, London, Special Publications. 2009;**323**(1):283-300

[69] Webb AA, Yin A, Harrison TM, Célérier J, Gehrels GE, Manning CE, et al. Cenozoic tectonic history of the Himachal Himalaya (northwestern India) and its constraints on the formation mechanism of the Himalayan orogen. Geosphere. 2011;**7**(4):1013-1061

[70] Pêcher A. Geology of the Nepal Himalaya: Deformation and petrography in the Main central Threust zone. Himalaya Colloquium of the International Association CNRS. 1977;**268**:301-308

[71] Bouchez JL, Pecher A. The Himalayan Main central thrust pile and its quartz-rich tectonites in Central Nepal. Tectonophysics. 1981;**78**(1-4): 23-50

[72] Brunel M. Ductile thrusting in the Himalayas: Shear sense criteria and stretching lineations. Tectonics. 1986;**5**(2):247-265

[73] Coward MP, Rex DC, Khan MA, Windley BF, Broughton RD, Luff IW, et al. Collision tectonics in the NW Himalayas. Geological Society, London, Special Publications. 1986;**19**(1):203-219

[74] Pêcher A. The metamorphism in central Himalaya, its relations with the thrust tectonic. Sciences de la Terre Memorial. 1986;**47**:285-309

[75] Searle MP. Structural evolution and sequence of thrusting in the high Himalayan, Tibetan—Tethys and Indus suture zones of Zanskar and Ladakh, Western Himalaya. Journal of Structural Geology. 1986;**8**(8):923-936

[76] Thakur VC. Plate tectonic interpretation of the western Himalaya. Tectonophysics. 1987;**134**(1-3):91-102

[77] Jain AK, Anand A. Deformational and strain patterns of an intracontinental collision ductile shear zone—An example from the higher Garhwal Himalaya. Journal of Structural Geology. 1988;**10**(7):717-734

[78] Jain AK, Manickavasagam RM. Inverted metamorphism in the intracontinental ductile shear zone during Himalayan collision tectonics. Geology. 1993;**21**(5):407-410

[79] Patel RC, Singh S, Asokan A, Manickavasagam RM, Jain AK. Extensional tectonics in the Himalayan orogen, Zanskar, NW India. Geological Society, London, Special Publications. 1993;**74**(1):445-459

[80] Grujic D, Casey M, Davidson C, Hollister LS, Kündig R, Pavlis T, et al.

Ductile extrusion of the higher Himalayan crystalline in Bhutan: Evidence from quartz microfabrics. Tectonophysics. 1996;**260**(1-3):21-43

[81] Spencer CJ, Harris RA, Dorais MJ. Depositional provenance of the Himalayan metamorphic core of Garhwal region, India: Constrained by U–Pb and Hf isotopes in zircons. Gondwana Research. 2012;**22**(1):26-35. DOI: 10.1016/j.gr.2011.10.004

[82] Cawood PA, Johnson MR, Nemchin AA. Early Palaeozoic orogenesis along the Indian margin of Gondwana: Tectonic response to Gondwana assembly. Earth and Planetary Science Letters. 2007;**255**(1-2):70-84

[83] Webb AA, Yin A, Dubey CS. U-Pb zircon geochronology of major lithologic units in the eastern Himalaya: Implications for the origin and assembly of Himalayan rocks. Bulletin. 2013 Mar 1;**125**(3-4):499-522. DOI: 10.1130/B30626.1

[84] Honegger K, Dietrich V, Frank W, Gansser A, Thöni M, Trommsdorff V. Magmatism and metamorphism in the Ladakh Himalayas (the Indus-Tsangpo suture zone). Earth and Planetary Science Letters. 1982;**60**(2):253-292

[85] Steck A, MatthieuGirard AM, Robyr M. Geological transect across the Tso Morari and Spiti areas: Thenappe structures of the Tethys Himalaya. Eclogae Geologicae Helvetiae. 1998;**91**:103-121

[86] DiPietro JA, Pogue KR. Tectonostratigraphic subdivisions of the Himalaya: A view from the west. Tectonics. 2004;**23**(5):119. DOI: 10.1029/2003TC001554

[87] Burchfiel BC, Zhiliang C, Hodges KV, Yuping L, Royden LH, Changrong D, et al. The south Tibetan detachment system, Himalayan orogen: Extension contemporaneous with and

parallel to shortening in a collisional mountain belt. Special Paper of the Geological Society of America. 1992;**267**:41

[88] Srivastava P, Mitra G. Thrust geometries and deep structure of the outer and lesser Himalaya, Kumaon and Garhwal (India): Implications for evolution of the Himalayan fold-and-thrust belt. Tectonics. 1994;**13**(1):89-109. DOI: 10.1029/93TC01130

[89] Dezes PJ, Vannay JC, Steck A, Bussy F, Cosca M. Synorogenic extension: Quantitative constraints on the age and displacement of the Zanskar shear zone (northwest Himalaya). Geological Society of America Bulletin. 1999;**111**(3):364-374

[90] Jackson M, Bilham R. Constraints on Himalayan deformation inferred from vertical velocity fields in Nepal and Tibet. Journal of Geophysical Research: Solid Earth. 1994;**99**(B7):13897-13912

[91] Dhiman R, Singh S. Neoproterozoic and Cambro-Ordovician magmatism: Episodic growth and reworking of continental crust, Himachal Himalaya. India. International Geology Review. 2021;**63**(4):422-436

[92] Medlicott HB. On the geological structure and relations of the Himalayan ranges, between the rivers Ganges and Ravee. Memoirs - Geological Survey of India. 1864;**3**:102

[93] Chaku SK. Geology of chauri tehsil and adjacent area, Chamba District. Himalayan Geology. 1972;**2**:405-414

[94] Bhatia GS, RC K. Dalhousie granites: Petrographic, petrochemical, Origin and emplacement aspects. Anais do Congresso Brasileiro de Geoquimica. 1989;**2**:64-65

[95] Le Fort P. The lower Paleozoic" lesser Himalayan" granitic belt: Emphasis on

the Simchar pluton of Central Nepal. Granites of Himalayas, Karakorum and Hindu Kush. 1983;**11**:235-255

[96] Schärer U, Xu RH, Allègre CJ. U (Th) Pb systematics and ages of Himalayan leucogranites, South Tibet. Earth and Planetary Science Letters. 1986;**77**(1):35-48

[97] Rolland Y, Mahéo G, Guillot S, Pêcher A. Tectono-metamorphic evolution of the Karakorum metamorphic complex (Dassu–Askole area, NE Pakistan): Exhumation of mid-crustal HT–MP gneisses in a convergent context. Journal of metamorphic Geology. 2001;**19**(6):717-737

[98] Islam R, Upadhyay R, Ahmad T, Thakur VC, Sinha AK. Pan-African magmatism, and sedimentation in the NW Himalaya. Gondwana Research. 1999;**2**(2):263-270

[99] Singh BN, Goel OP, Joshi M, Sheraton JW. Geochemistry and Petrogenesis of the Champawat Granitoids Occurring around Dhunaghat, District Pithoragarh, Uttar Pradesh, India. Journal of Geological Society of India (Online archive from Vol 1 to Vol 78). 1993;**42**(3):289-302

[100] Kumar S, Singh BN, Joshi M. Petrogenesis and Tectonomagmatic environment of Cambro-Ordovician granitoids of lesser Himalaya: A reappraisal. Visesa Prakasana-Bharatiya Bhuvaijñanika Sarveksana. 1996;**21**: 205-214

[101] Mehta PK. Rb-Sr geochronology of the Kulu-Mandi belt: Its implications for the Himalayan tectogenesis. Geologische Rundschau. 1977;**66**(1):156-175. DOI: 10.1007/BF01989570

[102] Miller C, Thöni M, Frank W, Grasemann B, Klötzli U, Guntli P, et al. The early Palaeozoic magmatic event in the Northwest Himalaya, India: source, tectonic setting and age of

emplacement. Geological Magazine. 2001 May;**138**(3):237-251. DOI: 10.1017/S0016756801005283

[103] Bikramaditya RK, Singh AK, Chung SL, Sharma R, Lee HY. Zircon U–Pb ages and Lu–Hf isotopes of metagranitoids from the Subansiri region, eastern Himalaya: Implications for crustal evolution along the northern Indian passive margin in the early Paleozoic. Geological Society, London, Special Publications. 2019;**481**(1):299-318

[104] Searle MP, Cottle JM, Streule MJ, Waters DJ. Crustal melt granites and migmatites along the Himalaya: Melt source, segregation, transport and granite emplacement mechanisms. Earth and Environmental Science Transactions of the Royal Society of Edinburgh. 2009;**100**(1-2):219-233

[105] Walker JD, Martin MW, Bowring SA, Searle MP, Waters DJ, Hodges KV. Metamorphism, melting, and extension: Age constraints from the high Himalayan slab of southeast Zanskar and northwest Lahaul. The Journal of Geology. 1999;**107**(4):473-495

[106] Huang C, Zhao Z, Li G, Zhu DC, Liu D, Shi Q. Leucogranites in Lhozag, southern Tibet: Implications for the tectonic evolution of the eastern Himalaya. Lithos. 2017;**294**:246-262

[107] Patel RC, Adlakha V, Lal N, Singh P, Kumar Y. Spatiotemporal variation in exhumation of the Crystallines in the NW-Himalaya, India: Constraints from fission track dating analysis. Tectonophysics. 2011;**504**(1-4):1-3

[108] Baud AY, Gaetani M, Garzanti E, Fois E, Nicora W, Tintori A. Geological observation in southeastern Zanskar and adjacent Lahul area (northern Himalaya). Eclogae Geologicae Helvetiae. 1984;**77**(1):177-197

[109] Garzanti E, Casnedi R, Jadoul F. Sedimentary evidence of a

Cambro-Ordovician orogenic event in the northwestern Himalaya. Sedimentary Geology. 1986;**48**(3-4):237-265

[110] Garzanti E, Baud A, Mascle G. Sedimentary record of the northward flight of India and its collision with Eurasia (Ladakh Himalaya, India). Geodinamica Acta. 1987;**1**(4-5):297-312

[111] Gaetani M, Garzanti E. Multicyclic history of the northern India continental margin (northwestern Himalaya). AAPG bulletin. 1991;**75**(9):1427-1446

[112] Brookfield ME. The Himalayan passive margin from Precambrian to cretaceous times. Sedimentary Geology. 1993;**84**(1-4):1-35

[113] Garzanti E. Sedimentary evolution and drowning of a passive margin shelf (Giumal group; Zanskar Tethys Himalaya, India): Palaeoenvironmental changes during final break-up of Gondwanaland. Geological Society, London, Special Publications. 1993;**74**(1):277-298

[114] Garzanti E. Stratigraphy and sedimentary history of the Nepal Tethys Himalaya passive margin. Journal of Asian Earth Sciences. 1999;**17**(5-6): 805-827

[115] Steck A, Spring L, Vannay JC, Masson H, Bucher H, Stutz E, et al. The tectonic evolution of the northwestern Himalaya in eastern Ladakh and Lahul, India. Geological Society, London, Special Publications. 1993;**74**(1):265-276

[116] Critelli S, Garzanti E. Provenance of the lower tertiary Murree redbeds (Hazara-Kashmir syntaxis, Pakistan) and initial rising of the Himalayas. Sedimentary Geology. 1994;**89**(3-4): 265-284

[117] Liu G, Einsele GJ. Sedimentary history of the Tethyan basin in the Tibetan Himalayas. Geologische Rundschau. 1994;**83**(1):32-61

[118] Liu G, Einsele G. Jurassic sedimentary facies and paleogeography of the former Indian passive margin in southern Tibet. Special Papers-Geological Society of America. 1999;**1**:75-108

[119] Bhargava ON, Singh BP. Geological evolution of the Tethys Himalaya. Episodes. 2020;**43**(1):404-416

[120] Herren E. Zanskar shear zone: Northeast-southwest extension within the higher Himalayas (Ladakh, India). Geology. 1987;**15**(5):409-413

[121] Jain AK, Kumar D, Singh S, Kumar A, Lal N. Timing, quantification and tectonic modelling of Pliocene–quaternary movements in the NW Himalaya: Evidence from fission track dating. Earth and Planetary Science Letters. 2000;**179**(3-4):437-451

[122] Valdiya KS. Geology and Natural Environment of Nainital Hills. Gyānodaya Prakāshan: Kumaun Himalaya; 1988

[123] Carosi R, Lombardo B, Musumeci G, Pertusati PC. Geology of the higher Himalayan crystallines in Khumbu Himal (eastern Nepal). Journal of Asian Earth Sciences. 1999;**17**(5-6):785-803

[124] Rameshwar Rao D, Sharma R. Arc magmatism in eastern Kumaun Himalaya, India: A study based on geochemistry of granitoid rocks. Island Arc. 2011;**20**(4):500-519

[125] Phukon P, Sen K, Srivastava HB, Singhal S, Sen A. U-Pb geochronology and geochemistry from the Kumaun Himalaya, NW India, reveal Paleoproterozoic arc magmatism related to formation of the Columbia supercontinent. Bulletin. 2018;**130**(7-8): 1164-1176

[126] Rogers JJ, Santosh M. Configuration of Columbia, a Mesoproterozoic

supercontinent. Gondwana Research. 2002;**5**(1):5-22

[127] Hou G, Santosh M, Qian X, Lister GS, Li J. Configuration of the late Paleoproterozoic supercontinent Columbia: Insights from radiating mafic dyke swarms. Gondwana Research. 2008;**14**(3):395-409

[128] Kaur P, Zeh A, Chaudhri N, Gerdes A, Okrusch M. Nature of magmatism and sedimentation at a Columbia active margin: Insights from combined U–Pb and Lu–Hf isotope data of detrital zircons from NW India. Gondwana Research. 2013;**23**(3): 1040-1052

[129] Sakai H, Iwano H, Danhara T, Takigami Y, Rai SM, Upreti BN, et al. Rift-related origin of the P aleoproterozoic Kuncha F ormation, and cooling history of the K uncha nappe and T aplejung granites, eastern Nepal lesser Himalaya: A multichronological approach. Island Arc. 2013;**22**(3):338-360

[130] Larson K, Piercey S, Cottle J. Preservation of a Paleoproterozoic rifted margin in the Himalaya: Insight from the Ulleri-Phaplu-Melung orthogneiss. Geoscience Frontiers. 2019;**10**(3):873-883

[131] Hoffman PF, St Onge MR, Easton RM, Grotzinger J, Schulze DE. Syntectonic plutonism in north-central Wopmay orogen (early Proterozoic), Hepburn Lake map area, district of Mackenzie. Geological Survey of Canada. 1980;**80**(1A):171-177

[132] Hoffman PF, Bowring SA. Short-lived 1.9 Ga continental margin and its destruction, Wopmay orogen, Northwest Canada. Geology. 1984;**12**(2):68-72

[133] Meert JG, Torsvik TH. The making and unmaking of a supercontinent: Rodinia revisited. Tectonophysics. 2003;**375**(1-4):261-288

[134] DiPietro JA, Isachsen CE. U-Pb zircon ages from the Indian plate in Northwest Pakistan and their significance to Himalayan and pre-Himalayan geologic history. Tectonics. 2001;**20**(4):510-525

[135] Ahmad I, Khan S, Lapen T, Burke K, Jehan N. Isotopic ages for alkaline igneous rocks, including a 26 Ma ignimbrite, from the Peshawar plain of northern Pakistan and their tectonic implications. Journal of Asian Earth Sciences. 2013;**62**:414-424

[136] Ding H, Zhang Z. Neoproterozoic granitoids in the eastern Himalayan orogen and their tectonic implications. Precambrian Research. 2016;**285**:1-9

[137] Richards A, Parrish R, Harris N, Argles T, Zhang L. Correlation of lithotectonic units across the eastern Himalaya, Bhutan. Geology. 2006;**34**(5): 341-344

[138] Hoffman PF. Did the breakout of Laurentia turn Gondwanaland inside-out? Science. 1991;**252**(5011): 1409-1412

[139] Li ZX, Bogdanova S, Collins AS, Davidson A, De Waele B, Ernst RE, et al. Assembly, configuration, and break-up history of Rodinia: A synthesis. Precambrian Research. 2008;**160**(1-2):179-210

[140] Rogers JJ, Santosh M. Continents and Supercontinents. Oxford: Oxford University Press; 2004

[141] Dalziel IW. OVERVIEW: Neoproterozoic-Paleozoic geography and tectonics: Review, hypothesis, environmental speculation. Geological Society of America Bulletin. 1997;**109**(1):16-42

[142] Unrung R. Rodinia to Gondwana: The Geodynamic map of Gondwana Supercontinent Assembly. Geological Society of America Bulletin Today. 1997;**7**(1):1-6

[143] Meert JG, Van Der Voo R. The assembly of Gondwana 800-550 Ma. Journal of Geodynamics. 1997;**23**(3-4): 223-235

[144] Oinam G, Singh AK, Joshi M, Dutt A, Singh MR, Singh NL, et al. Continental extension of northern Gondwana margin in the eastern Himalaya: Constraints from geochemistry and U–Pb zircon ages of mafic intrusives in the Siang window, Arunachal Himalaya, India. Comptes Rendus Geoscience. 2020;**352**(1):19-41

[145] Palin RM, Treloar PJ, Searle MP, Wald T, White RW, Mertz-Kraus R. U-Pb monazite ages from the Pakistan Himalaya record pre-Himalayan Ordovician orogeny and Permian continental breakup. Bulletin. 2018;**130**(11-12):2047-2061

[146] Bhargava ON, Thöni M, Miller C. Isotopic evidence of early palaeozoic metamorphism in the lesser Himalaya (Jutogh group), Himachal Pradesh, India: its implication. Himalayan Geology. 2016;**37**(2):73-84

[147] Argles TW, Prince CI, Foster GL, Vance D. New garnets for old? Cautionary tales from young mountain belts. Earth and Planetary Science Letters. 1999;**172**(3-4):301-309

[148] Foster GL. The pre-Neogene Thermal History of the Nanga Parbat Haramosh Massif and the NW Himalaya (Doctoral Dissertation). Milton Keynes, Buckinghamshire, United Kingdom: The Open University; 2016

[149] Catlos EJ, Sorensen SS, Harrison TM. Th-Pb ion-microprobe dating of allanite. American Mineralogist. 2000;**85**(5-6):633-648

[150] Catlos EJ, Harrison TM, Manning CE, Grove M, Rai SM, Hubbard MS, et al. Records of the evolution of the Himalayan orogen from in situ Th–Pb ion microprobe dating of monazite: Eastern Nepal and western Garhwal. Journal of Asian Earth Sciences. 2002;**20**(5):459-479

[151] Singh S, Barley ME, Brown SJ, Jain AK, Manickavasagam RM. SHRIMP U–Pb in zircon geochronology of the Chor granitoid: Evidence for Neoproterozoic magmatism in the lesser Himalayan granite belt of NW India. Precambrian Research. 2002;**118**(3-4): 285-292

[152] Zhou MF, Ma Y, Yan DP, Xia X, Zhao JH, Sun M. The Yanbian terrane (southern Sichuan Province, SW China): A Neoproterozoic arc assemblage in the western margin of the Yangtze block. Precambrian Research. 2006;**144**(1-2):19-38

[153] Gehrels GE, DeCelles PG, Martin A, Ojha TP, Pinhassi G, Upreti BN. Initiation of the Himalayan orogen as an early Paleozoic thin-skinned thrust belt. GSA Today. 2003;**13**(9):4-9

[154] Dézes P. Tectonic and Metamorphic Evolution of the Central Himalayan Domain in Southeast Zanskar (Kashmir, India) [Ph. D. thesis]. Lausanne, Switzerland: University of Lausanne; 1999

[155] Patriat P, Achache J. India–Eurasia collision chronology has implications for crustal shortening and driving mechanism of plates. Nature. 1984;**311**(5987):615-621

[156] Yeh MW, Shellnutt JG. The initial break-up of Pangæa elicited by late Palæozoic deglaciation. Scientific Reports. 2016;**6**(1):1-9

[157] Heuberger S, Schaltegger U, Burg JP, Villa IM, Frank M, Dawood H, et al. Age and isotopic constraints on magmatism along the Karakoram-Kohistan suture zone, NW Pakistan: Evidence for subduction and continued convergence after India-Asia collision.

Swiss Journal of Geosciences. 2007;**100**(1):85-107

[158] Angiolini L, Zanchi A, Zanchetta S, Nicora A, Vuolo I, Berra F, et al. From rift to drift in south Pamir (Tajikistan): Permian evolution of a Cimmerian terrane. Journal of Asian Earth Sciences. 2015;**102**:146-169

[159] Angiolini L, Zanchi A, Zanchetta S, Nicora A, Vezzoli G. The Cimmerian geopuzzle: New data from south Pamir. Terra Nova. 2013;**25**(5):352-360

[160] Pundir S, Adlakha V, Kumar S, Singhal S. Closure of India–Asia collision margin along the Shyok suture zone in the eastern Karakoram: New geochemical and zircon U–Pb geochronological observations. Geological Magazine. 2020;**157**(9):1451-1472

[161] Pundir S, Adlakha V, Kumar S, Singhal S, Sen K. Petrology, geochemistry and geochronology of granites and granite gneisses in the SE Karakoram, India: Record of subduction-related and pre-to syn-kinematic magmatism in the Karakoram fault zone. Mineralogy and Petrology. 2020;**114**(5):413-434

[162] Jain AK, Singh S. Tectonics of the southern Asian plate margin along the Karakoram shear zone: Constraints from field observations and U–Pb SHRIMP ages. Tectonophysics. 2008;**451**(1-4):186-205

[163] Kumar R, Jain AK, Lal N, Singh S. Early–middle Eocene exhumation of the trans-Himalayan Ladakh batholith, and the India–Asia convergence. Current Science. 2017;**25**:1090-1098

[164] Ravikant V, Wu FY, Ji WQ. Zircon U–Pb and Hf isotopic constraints on petrogenesis of the cretaceous–tertiary granites in eastern Karakoram and Ladakh, India. Lithos. 2009;**110**(1-4):153-166

[165] Upadhyay R. Implications of U–Pb zircon age of the Tirit granitoids on the closure of the Shyok suture zone, northern Ladakh, India. Current Science. 2008;**25**:1635-1640

[166] Thakur VC, Virdi NS, Rai H, Gupta KR. A note on the geology of Nubra-Shyok area of Ladakh, Kashmir, Himalaya. Journal of Geological Society of India (Online archive from Vol 1 to Vol 78). 1981;**22**(1):46-50

[167] Thakur VC. Tectonostratigraphic framework and the closing of neo-Tethys in Ladakh, NW trans Himalaya, India: A Review. Himalayan Geology. 2022;**43**(1B):201-220

[168] Borneman NL, Hodges KV, Van Soest MC, Bohon W, Wartho JA, Cronk SS, et al. Age and structure of the Shyok suture in the Ladakh region of northwestern India: Implications for slip on the Karakoram fault system. Tectonics. 2015;**34**(10):2011-2033

[169] Dupont-Nivet G, Lippert PC, Van Hinsbergen DJ, Meijers MJ, Kapp P. Palaeolatitude and age of the Indo–Asia collision: Palaeomagnetic constraints. Geophysical Journal International. 2010;**182**(3):1189-1198

[170] Dupont-Nivet G, Van Hinsbergen DJ, Torsvik TH. Persistently low Asian paleolatitudes: Implications for the India-Asia collision history. Tectonics. 2010;**29**(5):10562

[171] Adlakha V, Patel RC, Kumar A, Lal N. Tectonic control over exhumation in the Arunachal Himalaya: New constraints from apatite fission track analysis. Geological Society, London, Special Publications. 2019;**481**(1):65-79

[172] Godin L, Grujic D, Law RD, Searle MP. Channel flow, ductile extrusion and exhumation in continental collision zones: An introduction. Geological Society,

London, Special Publications. 2006;
268(1):1-23

[173] Montomoli C, Carosi R, Iaccarino S. Tectonometamorphic discontinuities in the greater Himalayan sequence: A local or a regional feature? Geological Society, London, Special Publications. 2015;**412**(1):25-41

[174] Tobgay T, McQuarrie N, Long S, Kohn MJ, Corrie SL. The age and rate of displacement along the Main central thrust in the western Bhutan Himalaya. Earth and Planetary Science Letters. 2012;**319**:146-158

[175] Hubbard MS, Harrison TM. 40Ar/39Ar age constraints on deformation and metamorphism in the Main central thrust zone and Tibetan slab, eastern Nepal Himalaya. Tectonics. 1989;**8**(4):865-880

[176] Vannay JC, Grasemann B. Himalayan inverted metamorphism and syn-convergence extension as a consequence of a general shear extrusion. Geological Magazine. 2001;**138**(3):253-276

[177] Catlos EJ, Harrison TM, Kohn MJ, Grove M, Ryerson FJ, Manning CE, et al. Geochronologic and thermobarometric constraints on the evolution of the Main central thrust, Central Nepal Himalaya. Journal of Geophysical Research: Solid Earth. 2001;**106**(B8):16177-16204

[178] Vannay JC, Grasemann B, Rahn M, Frank W, Carter A, Baudraz V, et al. Miocene to Holocene exhumation of metamorphic crustal wedges in the NW Himalaya: Evidence for tectonic extrusion coupled to fluvial erosion. Tectonics. 2004;**23**(1):TC1014

[179] Montemagni C, Fulignati P, Iaccarino S, Marianelli P, Montomoli C, Sbrana A. Deformation and fluid flow in the Munsiari Thrust (NW India): A preliminary fluid inclusion study.

Journal of the Geological Society of India. 2016;**82**:67-77

[180] Iaccarino S, Montomoli C, Carosi R, Massonne HJ, Visonà D. Geology and tectono-metamorphic evolution of the Himalayan metamorphic core: Insights from the Mugu Karnali transect, Western Nepal (central Himalaya). Journal of Metamorphic Geology. 2017;**35**(3):301-325

[181] Jain AK. Large-scale tectonic models for the evolution of the Himalaya. Himalayan Geology. 2022;**43**(1B):164-179

[182] Pebam J, Adlakha V, Jain AK, Patel RC, Lal N, Singh S, et al. Apatite and zircon fission-track thermochronology constraining the interplay between tectonics, topography and exhumation, Arunachal Himalaya. Journal of Earth System Science. 2021;**130**(3):1-21

[183] Meigs AJ, Burbank DW, Beck RA. Middle-late Miocene (> 10 Ma) formation of the Main boundary thrust in the western Himalaya. Geology. 1995;**23**(5):423-426

[184] Adlakha V, Patel RC, Lal N, Mehta YP, Jain AK, Kumar A. Tectonics and climate interplay: Exhumation patterns of the Dhauladhar range, northwest Himalaya. Current Science. 2013;**10**:1551-1559

[185] Singh P, Patel RC. Miocene development of the Main boundary thrust and Ramgarh thrust, and exhumation of lesser Himalayan rocks of the Kumaun-Garhwal region, NW-Himalaya (India): Insights from fission track thermochronology. Journal of Asian Earth Sciences. 2022;**224**:104987

[186] Hodges KV, Wobus C, Ruhl K, Schildgen T, Whipple K. Quaternary deformation, river steepening, and heavy precipitation at the front of the

higher Himalayan ranges. Earth and Planetary Science Letters. 2004; **220**(3-4):379-389

[187] Patel RC, Carter A. Exhumation history of the Higher Himalayan Crystalline along Dhauliganga-Goriganga river valleys, NW India: New constraints from fission track analysis. Tectonics. 2009;**28**(3):16

[188] Adlakha V, Lang KA, Patel RC, Lal N, Huntington KW. Rapid long-term erosion in the rain shadow of the Shillong plateau. Eastern Himalaya. Tectonophysics. 2013;**582**:76-83

[189] Burchfiel BC, Royden LH. North-south extension within the convergent Himalayan region. Geology. 1985;**13**(10):679-682

[190] Webb AA, Yin A, Harrison TM, Célérier J, Burgess WP. The leading edge of the greater Himalayan crystalline complex revealed in the NW Indian Himalaya: Implications for the evolution of the Himalayan orogen. Geology. 2007;**35**(10):955-958

Chapter 2

The Breaking of the Iranian Block during the Cretaceous and the Opening of New Oceanic Basins within the Tethys Ocean: The Case of the Sabzevar-Nain Basin and Its Geodynamic History

Saidi Abdollah, Khan Nazer Nasser, Hadi Pourjamali Zahra and Farzad Kiana

Abstract

The Jurassic subduction of the Neo-Tethys oceanic crust under the western continental margin of the Iranian Block has led to the fragmentation of the Iranian Block in the back-arc basin, leading to the opening of three oceanic basins around it. The ophiolitic belts surrounding central Iran are the indicators of the closure of these basins. The Sabzevar-Nain Basin is one of these basins, which has been created between the micro-block of central Iran in the south and the Alborz Mountain Ranges in the north. This basin opened in the late Jurassic as a rift and then became a trough in the early Cretaceous. Finally, this basin developed into an oceanic basin in the early late Cretaceous. The sedimentation in this basin can be divided into pre-rift, syn-rift and oceanic environments. All of these sediments are strongly folded and faulted. The closure of this basin started during the Paleocene with a subduction under the southern margin of the Alborz Mountain Ranges. The collision event between the northern margin of the micro-block of central Iran and the southern margin of the Alborz Mountain Ranges occurred in the early Eocene. The result of this event was the creation of a wide collision zone, forming a volcanic arc and a back arc basin on the active of the Alborz Mountain Ranges, an ophiolitic belt, and post- collision intrusion masses that appear everywhere in the collision zone. In the point of lithology, these intrusion masses are composed of granite, diorite, and granodiorite. The magmatic activities that started in the Paleocene-early Eocene continued until early Quaternary.

Keywords: geodynamics, ophiolite, collision, volcanic arc, syn-rift, foreland basin, back-arc basin

1. Introduction

It is possible to express the evolution of the Iranian crust and basins by looking at their geology and geodynamic history: the consolidation of the Iranian basement in the Gondwana mega continent; the Precambrian magmatism and metamorphism; deformation and folding in the Precambrian and early Paleozoic rocks; crustal thinning due to an extensional state during the early Paleozoic (Neo-Tethys rifting); rifting in the eastern part of Gondwana in the early Paleozoic [Silurian], (**Figure 1a**); the northward subduction of the paleo-Tethys oceanic crust under the southern margin of the Eurasia supercontinent in the early Paleozoic (**Figure 1b**); the early Cimmerian collision of the Iranian Block with the Turan Block (the southern margin of Eurasia) in the middle Paleozoic (**Figure 1c**); the late Paleozoic-early Jurassic crustal thinning with an instability period in the sedimentary basin after the early Cimmerian orogenic event (**Figure 1d**). The great event on the Iranian Block occurred after the oceanic crust subduction of the Neo- Tethys Ocean under the western margin of the Iranian Block. This subduction led to the global extension and fragmentation of the Iranian crust during the late Jurassic. After this extension and thinning, the crust of the Iranian Block broke in the back-arc basin of the Neo-Tethys Ocean in the early-middle Cretaceous. This event led to the creation of inner oceanic basins such as the Sabzevar-Nain, Nain-Baft, and Sistan-Baluch Basins (**Figure 2**) [1–5]. The ophiolitic belts surrounding central Iran are the indicators of the closure of these basins. They were over thrust on the continental margins accompanied by other materials of the accretionary prism and the continent-continent collision between the adjacent blocks.

In this paper, the authors tried to discuss the geochronology of the opening and closing processes of the Sabzevar-Nain Basin situated between the southern margin of the Alborz Mountain Ranges and the northern margin of

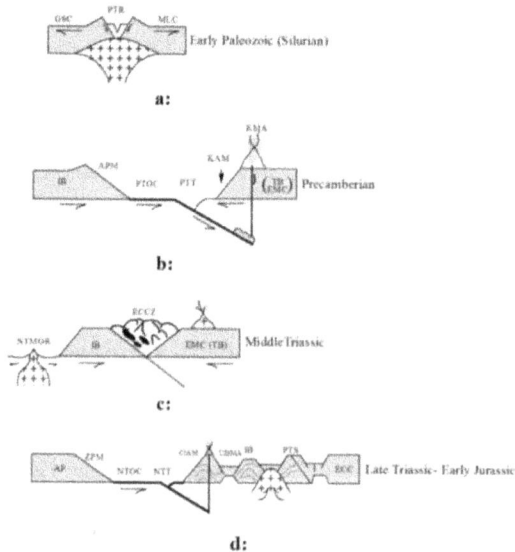

Figure 1.
A brief history of the Iranian block evolution from the Precambrian to the late Triassic-early Jurassic; a: The rifting event in the eastern part of Gondwana; b: The subduction of the paleo-Tethys slab under the Turan block (the southern part of Eurasia); c: The early Cimmerian collision of the Iranian block with the Eurasia mega-continent and the development of the neo-Tethys Ocean; d: The crustal thinning and instability in the sedimentary basin after the early Cimmerian collision. GSC: Gondwana supercontinent; PTR: Paleo-Tethys rift; MLC: Mega Lhasa continent; IB: Iranian block; APM: Alborz passive margin; PTOC: Paleo-Tethys oceanic crust; PTT: Paleo-Tethys trench; KMA: Kopeh Dagh magmatic arc; EMC (TB): Eurasia mega-continent (the Turan block); ECCZ: Early Cimmerian collision zone; AP: Arabian plate; ZPM: Zagros passive margin; NTOC: Neo-Tethys oceanic crust; NTT: Neo-Tethys trench; PTS: Paleo-Tethys suture; ECC: Eurasia continental crust.

the micro-block of central Iran (**Figure 3**). The Sabzevar-Nain Basin (SNB) was formed due to an extensional system with crustal thinning accompanied by a north-south rifting which occurred in the back-arc basin of the Neo-Tethys Ocean. The large number of geological structures in this region has motivated many geologists to conduct their researches in it. Several Ph.D. dissertations [including Sadreddini [6], Alavi Tehrani [7], Dehghani [8], and Noghreyan [9]] have been conducted on the oceanic crust remnants (ophiolites) in the Sabzevar region. The first study on the Sabzevar ophiolites was conducted by Sadreddini [6]. This study depended on the petrographic characteristics of the ophiolites in the middle part of the ophiolitic range of Sabzevar. Alavi Tehrani [7] studied the geology and petrology of the ophiolitic rocks in the northwest of Sabzevar. Dehghani [8] described the gravity field and structure of the Iranian crust. Noghreyan [9] stated that the ophiolitic belt of Sabzevar was formed due to an immature arc. These studies were carried out in the frame of the Geological Survey of Iran called the 'Geodynamic Project, [10–12]. The ophiolitic belt was distinguished by rock units including harzburgite, intrusive rocks, a sheeted dyke complex, volcano-sedimentary sequences, an ophiolitic mélange (composed of ophiolitic rocks, intrusive rocks, and oceanic sediments), and metamorphic rocks. The cover rocks were characterized by a Cenozoic sequence and Quaternary rocks. Lensch and Davoudzadeh [13] identified three types of ophiolitic rocks around the micro-block of central Iran including an ophiolitic mélange, ridge type ophiolites, and trench type ophiolites. Baghdadi [14] related the volcanism of the northern part of Sabzevar to a subduction process during

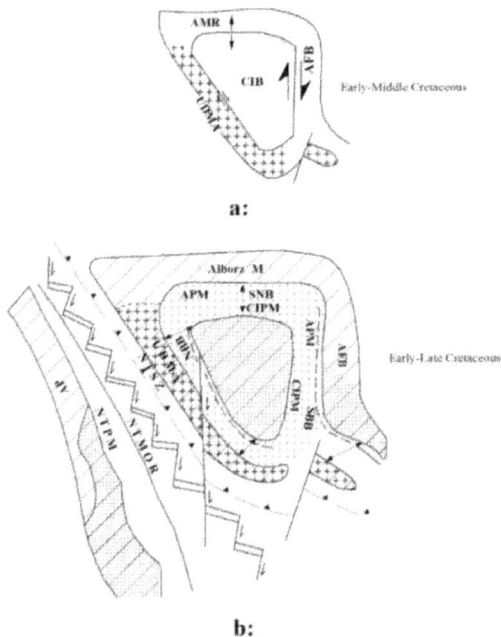

Figure 2.
The breaking of the Iranian block and the creation of rifts around the micro-block of Central Iran (a), the locations of the three new basins around the micro-block of Central Iran in the back-arc of the neo-Tethys Ocean (b). AMR: Alborz Mountain ranges; AFB: Afghan block; CIB: Central Iran block; UDMA: Urumieh-Dokhtar magmatic arc; Alborz M.: Alborz Mountains; APM: Alborz passive margin; CIPM: Central Iran passive margin; NBB: Nain-Baft Basin; SNB: Sabzevar-Nain Basin; SBB: Sistan- Baluch Basin; APM: Afghan passive margin; AP: Arabian plate; NTMP: Neo-Tethys passive margin; NTMOR: Neo-Tethys mid-oceanic ridge; NTSZ: Neo-Tethys suture zone.

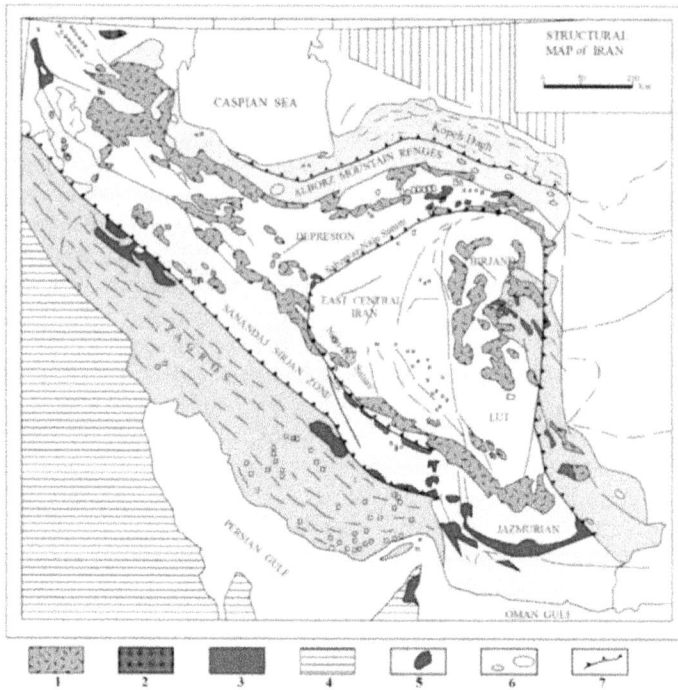

Figure 3.
The location of the oceanic crust remnants and the suture of the Sabzevar collision zone in the northern part of the micro-block of Central Iran: 1. Cenozoic volcanic rocks; 2. The intrusion of Mesozoic and Cenozoic granites and diorites; 3. The cretaceous-Paleocene ophiolites around the micro-block of Central Iran; 4. The ophiolites of the Zagros suture zone; 5. The ophiolites of the paleo-Tethys continental collision exposed in the north of Iran; 6: The Hormoz formation (pre- Cambrian) in the southeast of the Zagros Mountain ranges; 7: The trace of the paleo-Tethys suture.

Eocene. Ghassemi and Rezaei-Kahkhaei [15] stated that these rocks were the result of the partial melting of an enriched mantle by an extensional process within the arc. Khalatbari and Etessami [16] worked on the petrology and tectonomagmatic setting of the Eocene volcanic rocks in the Semnan area. They concluded that these volcanic rocks are the result of a subduction event during the Paleocene. Based on her studies on the petrology, petrography, and geochemistry of the intrusive rocks of the Sabzevar-Nain collision zone, Goharshahi [17] concluded that their exposure is due to a subduction event and its continuation after the collision (post-collision intrusion).

During the years 1999–2002, 12 geological maps in the scale of 1:100,000 were prepared in the framework of a project in the Geological Survey of Iran. They contain important data about the petrology, stratigraphy, sedimentology, and structural geology of the geodynamic events.

2. Geological setting

The geological characteristics of the Sabzevar-Nain collision zone are described as pre- and syn-rift sediments, syn-rift magmatic activities, syn-subduction oceanic sediments, syn-subduction magmatism, syn-collision volcanism, post-collision sediments, post-collision magmatism, and post-collision volcanism. The pre-rift sediments are a thick Paleozoic and Mesozoic sequence including the sediments of continental and platform environments. The youngest pre-rift sediments of this basin are late Jurassic-early Cretaceous carbonates which have a wide distribution in the southern part of the Alborz Mountains and the northern part of central Iran (Saidi and Akbarpour [18]; Saidi and Vahdati Daneshmand [19]). Their thickness changes from 450 m in the north of Damghan to 580 m in the south of Sabzevar.

The syn-rift sequence is composed of continental and detrital facies as well as thick-bedded and massive limestone of the platform environment (**Figure 4**). The other syn-rift sediments are thick highly deformed flysch facies consisting of a majority of calcareous shale, some sandstone, and a few limestone lenses. These facies show the high thinning of the continental crust at the time of sedimentation in the Sabzevar-Nain trough. The basalts in the magma chamber under the rift penetrated the broken crust and flowed on the basin floor. In the convergent system between the continental crusts of the Alborz Mountain Ranges and central Iran, these basalts were folded with the deposited sediments in the basin (Salamati and Shafei [20]; Kolivand [21]; Ghaffari Nik [22]) (**Figure 5**).

Figure 4.
The two different facies of middle Cretaceous in the middle part of collision zone. The low lands are the syn-rift sediments derived from the continent (flysch) which are covered by upthrusting the thick bedded, massive limestone of platform environments (high lands).

Figure 5.
The physical characteristics of the syn-rift sediment during middle Cretaceous with intercalation of basalt.

The pre- and syn-collision late Cretaceous oceanic deposits consist of shale, sandstone, tuffaceous shale, limestone, radiolarian shale, pelagic limestone (Campanian-Maastrichtian in age), tuff, pillow lava, and spilitic basalt. The thickness of these sediments is estimated to be 2000 m (**Figure 6**) [23].

The syn-subduction and syn-collision Paleocene-Eocene volcanic rocks of the volcanic arc in the Sabzevar region are composed of andesite, andesitic tuff, andesitic basalt, olivine alkali breccia, feeder dykes, basaltic andesite, and porphyritic andesite (**Figures** 7 and **8**) [24].

The post-collision foreland basin deposits consist of conglomerate, sandy limestone, gypsiferous marl, tuffaceous shale, and sandstone (Eocene and Oligo-Miocene in age). The other sediments of this basin are Miocene sandstone, conglomerate, and gypsiferous marl (**Figures 9** and **10**).

The syn/post-collision intrusive rocks consist of Cenozoic granite, quartz diorite, and diorite (**Figures 11** and **12**).

Figure 6.
The late Cretaceous oceanic deposits consist of shale, sandstone, pelagic limestone, radiolarian shale, pillow lava (Campanian- Maastrichtian).

Figure 7.
The morphology characteristics of the Paleocene-Eocene volcanic arc in the western part of Abbasabad (west of Sabzevar).

Figure 8.
The syn-subduction and syn-collision Paleoceane-Eoceane volcanic arc parallel with the ophiolitic belt, north of Sabzevar.

Figure 9.
The brown well-bedded post collision foreland deposits, north west of Sabzevar, composed of conglomerate and sandstone.

Figure 10.
The dark brown thick bedded post collision foreland basin deposits, over the ophiolitic rocks of Sabzevar-Nain Suture (East of Davarzan).

Figure 11.
The great masses of syn/post collision intrusive rocks, consist of Cenozoic granitoid, diorite and granodiorite in Kuh-e-Baharestan (south of Sheshtamad).

Figure 12.
The granite masses of syn/post collision, in age of Paleocene in the Kuh-e-Mish (south of Sabzevar).

The most recent post-collision dacite domes (Miocene and Plio-Quaternary in age) intruded into the ophiolitic belt (**Figure 13**), andesitic lava, tuff, and dacitic lava flow of Sabzevar (**Figure 14**).

3. Geodynamic setting

The architecture of the present-day Sabzevar-Nain Basin reflects the extensive tectonic regime (**Figure 2a**) which has occurred since the early-middle Cretaceous [2, 4]. This basin has been formed along the east-west direction in the southern part of the Alborz Mountain Ranges and the northern micro-block of central Iran. The tectonostratigraphic sequences of the sediments deposited in the basin between the late Paleozoic and Cenozoic are shown in **Figure 15**. These sediments are composed of shallow water deposits of early Triassic before early Cimmerian orogenic events (the continent-continent collision of the Iranian Block and Eurasia) (**Figure 1c**).

Figure 13.
Youngest post collision dacitic domes penetrated into the ophiolitic rocks of Sabzevar.

Figure 14.
Young post collision dacitic domes penetrated into the foreland basin deposits, north of Mehr (North-West of Sabzevar).

The early Triassic deposits are unconformably overlain by shale and sandstone detrital facies with an age of late Triassic-early Jurassic and from a new basin in extension after the above-mentioned orogenic events. A discontinuity can be observed in the sedimentary sequences between the late Jurassic and early-middle Cretaceous deposits (Late Cimmerian events). The first stage of rifting occurred in the early-middle Cretaceous in the basins around the central Iranian Block (**Figure 2b**) which were mechanically different from the Nain-Baft and Sistan-Baluch basins. The last two basins were created as pull-apart basins along two transform faults in the western and eastern parts of the micro-block of central Iran. The Sabzevar-Nain Basin was formed in the form of a classical continental rift (**Figure 16a**). This was the time for the micro-block of central Iran to be separated from the Alborz Mountain Ranges and to move toward the south.

During the middle Cretaceous, when a magma chamber was formed below the rift of Sabzevar-Nain, this basin became a large but not very wide trough (**Figure 16b**). After the process of intercontinental extension and the formation

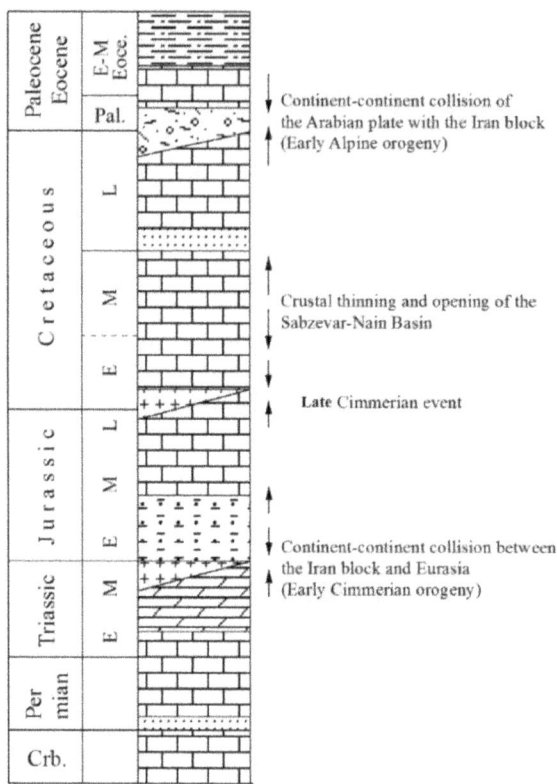

Figure 15.
The tectonostratigraphic sequences of the Iranian block before the middle Cretaceous segmentation.

of the oceanic crust, the Sabzevar-Nain trough changed to an oceanic basin
(**Figure 16c**). Simultaneously with the change of the divergent regime to the
convergent regime during the Santonian-Campanian (86.3 ± 0.5–72.1 ± 0.2 Ma)
[25], the oceanic crust of the Sabzevar-Nain Basin was subducted under the
Alborz continental crust (**Figure 16d**). At this time, sedimentation reached its
highest rate, whereas the intensity of deposition reached its highest rate in the
Maastrichtian. Therefore, in 6.1 Ma, more than 578 m of sediments were deposited in the basin [23].

The process of subduction beneath the Alborz continental crust continued
until the beginning of the early Eocene, while the margin of central Iran always
remained a passive margin (**Figure 16e**). Since the early Paleocene, there have been
highly intense volcanic and intrusive activities which can be observed throughout
the collision zone especially in the south of Sabzevar. The oldest age determined for
these volcanic and intrusive rocks is Paleocene [7, 9, 13, 14, 17, 26]. However, the
age of the volcanic activity in the collision zone and the back-arc of the Sabzevar-
Nain Basin varies from Paleocene to early Quaternary [20, 23, 24, 27–31]. The
convergence and closure of the Sabzevar-Nain Basin could be due to the northward
movement of the micro-block of central Iran. This event took place during the
late Eocene-Oligocene (**Figure 16f**). The youngest deposits in the Sabzevar-Nain
Basin are early Eocene flysch facies which were upthrust on both continental
margins. These sediments are widely spread in the Nain region and the northwest of
Bardeskan in the south of Sabzevar.

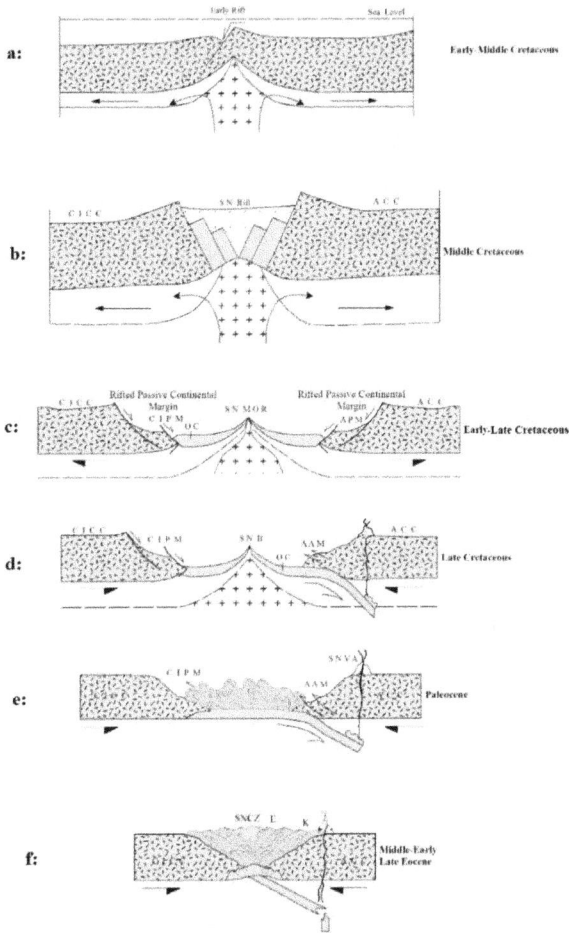

Figure 16.
The geodynamic modeling of the opening (early-middle Cretaceous) and closure (middle-early late Eocene) of the Sabzevar-Nain Basin. CICC: Central Iran continental crust; S N rift: Sabzevar-Nain rift; ACC: Alborz continental crust; CIPM: Central Iran passive margin; APM: Alborz passive margin; SNMOR: Sabzevar-Nain mid oceanic ridge; SNB: Sabzevar-Nain Ocean; SNVA: Sabzevar-Nain volcanic arc; SNCZ: Sabzevar-Nain collision zone; AAM: Alborz active margin; OC: Oceanic crust; E: Eocene; K: Cretaceous.

4. The structural and petrological characteristics of the Sabzevar-Nain collision zone

The best and the more complete units of the Sabzevar-Nain collision zone are well appeared around the Sabzevar. In a section from north to south, these units may be described as below:

Back arc basin in which, is deposited the detrital sediments in age of late Eocene, Oligo-Miocene and plio-Quaternary, Volcanic arc mainly composed of Eocene intermediate volcanic rocks and ophiolitic belt (Sabzevar-Nain oceanic crust remnants). Foreland basin in which the shallow water sediments of early- middle Eocene to Plio- Quaternary are accumulated., Syn-post collision intrusive masses of granite, granodiorite, diorite and granitoid, and overthrustihg of ophiolitic units (**Figure 17**) are also observed.

Figure 17.
The structural sections of the Sabzevar-Nain collision zone. SNBAB: Sabzevar-Nain back arc basin; SNVA: Sabzevar-Nain volcanic arc; EVSR: Eocene volcano- sedimentary rocks; SNFB: Sabzevar-Nain foreland basin; SNOCR: Sabzevar-Nain oceanic crust remnants (ophiolites); CIPM: Central Iran passive margin; AAM: Alborz active margin; JBB: Joghatay back arc basin; BFB: Bardeskan foreland basin; DF: Daruneh fault; PCDA: Post collision dacites; QVC: Quaternary volcanic Cone; EM: Eocene- Miocene deposits in foreland basin; K_2^l: Late Cretaceous oceanic deposits; E^v: Eocene volcanic; Q: Quaternary; M: Miocene deposits; E: Eocene foreland basin deposits; gr: Post collision granites; d: Post collision diorites.

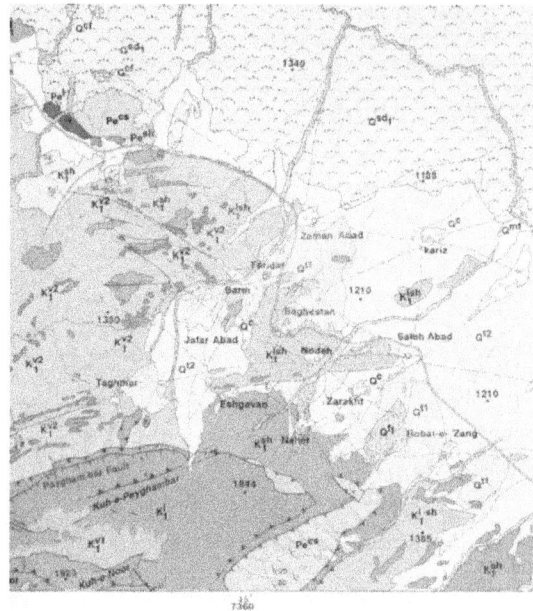

Scale: 1:200,000

Figure 18.
The syn-rift basaltic flows (K_1^{v2}) folded with the sediments of the Sabzevar-Nain trough. Quaternary: 1. Clay flat, 2. Sand dunes, 3. Cultivated area, 4. Mud flat, 5. Younger alluvium, 6. Older alluvium, 7. High level gravel fan, 8. Low level gravel fan, Cretaceous: 9. Crystal lithic tuff, 10. Andesitic lava (spilite), 11. Calcareous shale and limestone (turbidites), 12. Massive limestone, 13. Shaly limestone, 14. Shale, Paleocene: 15. Conglomerate and sandstone, 16. Shale.

Middle Cretaceous basalts: These rocks comprise the oldest layer of the Sabzevar-Nain oceanic crust. In fact, these basaltic flows are the highest part of the magma chamber under the Sabzevar-Nain rift. These rocks can be observed within the flysch sediments of the trough and are folded with them during continental convergence (**Figure 18**). These are from alkali basalt series and have a spilitic texture [20]. In the Kharturan area, the syn-rift sediments (flysch facies) and the massive carbonates of the platform environment have been exposed near each other (**Figure 4**).

Late Cretaceous deep-sea sediments: These rocks are widely exposed in the middle part of the collision zone especially in the north and southwest of

Kuh-e-Baharestan (Baharestan Mountain) in the south of Sabzevar. These sediments are composed of siliceous sandstone, shale and tuffaceous sandstone with some pillow lava, spilitic vesicular lava, green tuff, radiolarian shale, and pelagic limestone. These deep sediments are strongly folded and faulted. They are a combination of thrusting slices and their boundaries with other rock units are thrust faults.

Oceanic crust remnants: The ophiolitic rocks of the Sabzevar-Nain oceanic crust appear in three areas of the collision zone. The southern exposure is near the northern margin of central Iran among the Eocene flysch and volcanic rocks (**Figure 19**). Here, they are composed of ultramafic rocks, diabase, plagiogranite, and gabbro. The second one is exposed with a fault contact in the middle part of Kuh-e Mish just in the southern flank of the great post-collision exposures of diorite. At their southern limit, these ophiolitic rocks are upthrust on Miocene marls. Lithologically, they are composed of harzburgite, serpentinite, dunite, diabase, and gabbro (**Figure 20**) [7, 13, 23, 32]. The main remnants of the oceanic

Figure 19.
Southern exposure of ophiolitic rocks near the northern margin of Central Iran composed of ultramafic rocks, diabase, plagiogranite and gabbro.

Figure 20.
The second ophiolitic rocks are exposed by thrust faults in the middle part of Kuh- e- Mish. In the southern limit, they are upthrust on the Miocene marls.

crust in the suture zone (**Figure 17**) in the north of Sabzevar are exposed as a mountain range. The length of this mountain range is more than 480 km from east to west and its width is about 20 km. In its southern limits, this ophiolitic range is thrust over the detrital sediments of the foreland basin (**Figure 21**). However, in its northern part, it is limited to the volcanic arc (**Figure 17**). From a lithological point of view, it is composed of harzburgite, lherzolite, dunite, rodingite, serpentinized harzburgite, gabbro, diabase, a complex of sheeted dykes, submarine andesitic basalt, pillow lava, amphibolite, amphibolite schists, serpentinized peridotite, glaucophane schists, hornblende schist, garnet-muscovite schists, epidote-muscovite-chlorite schist, epidote-tremolite-actinolite schists, marble, pegmatite gabbro, monzodiorite, diorite, quartz diorite, and granodiorite [7, 9, 13, 24, 26–28, 33] (**Figure 22**). Obviously, the last intrusions are related to post-collision intrusion events (**Figure 23**) [17].

Volcanic arc: The Sabzevar-Nain volcanic arc is exposed on the Alborz continental margin very close to the suture zone and parallel to the ophiolitic range.

Figure 21.
The main remnant of oceanic crust in the suture zone in the north of Sabzevar. These ophiolitic rocks are thrusted over the foreland basin deposits.

Figure 22.
The ophiolitic rocks, situated in the northeast of Sabzevar, composed of harzburgite, lherzolite, dunite, rodengite, serpentinite, gabbro, diabase and pillow basaltic lava.

Figure 23.
The last intrusion related to syn- post collision activities (Kuh- e- Mish) (south of Sabzevar).

Figure 24.
The Sabzevar- Nain volcanic arc exposed on the Alborz continental margin close to the suture zone, mostly composed of intermediate volcanic rocks.

This can be due to the 30–35-degree angle of the oceanic crust subducted under the Alborz continental crust. Its length is about 660 km from the west of Torbat-e-Jam in the east of Iran to the north of Semnan in the central southern part of the Alborz Mountain Ranges. Its width in the west of Abbasabad near Miamey is a little more than 10 km. The lithological composition of this arc changes from east to west. In the Sabzevar region, it is reported as red dacite, pyroclastic rocks, massive micrite, and trachyandesite (Oligocene in age). The Eocene volcanic rocks are siliceous tuff interbedded with andesite, porphyritic andesite, basalt, agglomerate andesite basalt to trachyandesite basalt, volcanic breccia, andesite lava, and lithic crystal tuff (**Figure 24**) [23, 27, 28, 30]. In the western part of the arc in the north of Semnan, the volcanic rocks from bottom to top include intermediate-basic lava with a composition of andesite basalt to phitic andesite. In some parts, this sequence is crosscut by andesitic dykes. The eruptions of andesitic-basaltic lava sometimes enter the shallow water environment and produce brecciated hyaloclastites with sandstone, shale, and limestone. In the higher parts of the sequence, there is some intermediate lava with a phyric andesite composition which is crosscut by quartz feldspathic dykes. Briefly, the Eocene volcanic rocks in the western part of the volcanic arc consist of basalt, basaltic andesite, andesite, dacite, riodacite, riolite, and tuff. Based on their geological, petrographic, and geochemical study (2018), Khalatbari and Etessami concluded that these volcanic rocks are the result of a subduction event during the Paleocene and early-middle Eocene (**Figure 25**).

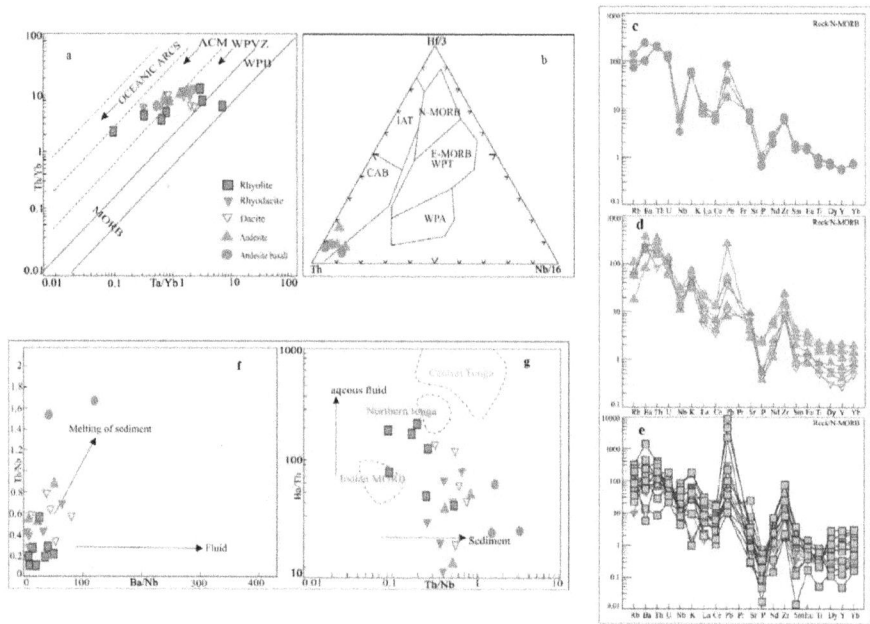

Figure 25.
Location of the studied samples on the diagrams determination of the tectonomagmatic environment a) Th-Hf/3-Nb/16 diagram [34]. b) Ta/Yb vs. Th/Yb diagram ([35]). (c,d,e) normalized multi- element spider diagrams with N- MORB value [36] for the Ahovan (Semnan) volcanic rock samples. Diagrams to investigate the role of subduction compositions (fluid/melting). f) Ba/Nb diagram vs. Th/Nb [37]. g) Th/Nb diagram vs. Ba/Th [38]. (after [16]).

In their geological, petrographic, and geochemical studies, Baghdadi [14] and Shahosseini and Ghassemi [39] reached the same conclusion regarding the north and west of Sabzevar, respectively. In their petrochemical and tectonic setting study of the Davarzan-Abassabad volcanic (DAEV) rocks, Ghassemi and Rezaei-Kahkhaei [15] stated that these volcanic rocks are the product of the partial melting of an enriched mantle by an extensional event within the arc (**Figures 26–29**).

Post-collision intrusive masses: The post-collision intrusive rocks are cropped out in the form of small and large masses everywhere in the collision zone. The intrusive masses within the ophiolitic belt in the north of Sabzevar are usually small scale. The greatest masses appear in the middle parts of the collision zone in the south and southeast of Sabzevar in the mountains called Borj-Kuh, Kuh-e-Mish, and Kuh-e-Baharestan. The petrological studies on the small intrusive bodies within the ophiolites have shown that they are composed of granite, granophyre, granitoid [29], granite, quartz diorite, diorite, microdiorite [31], micromonzonite,

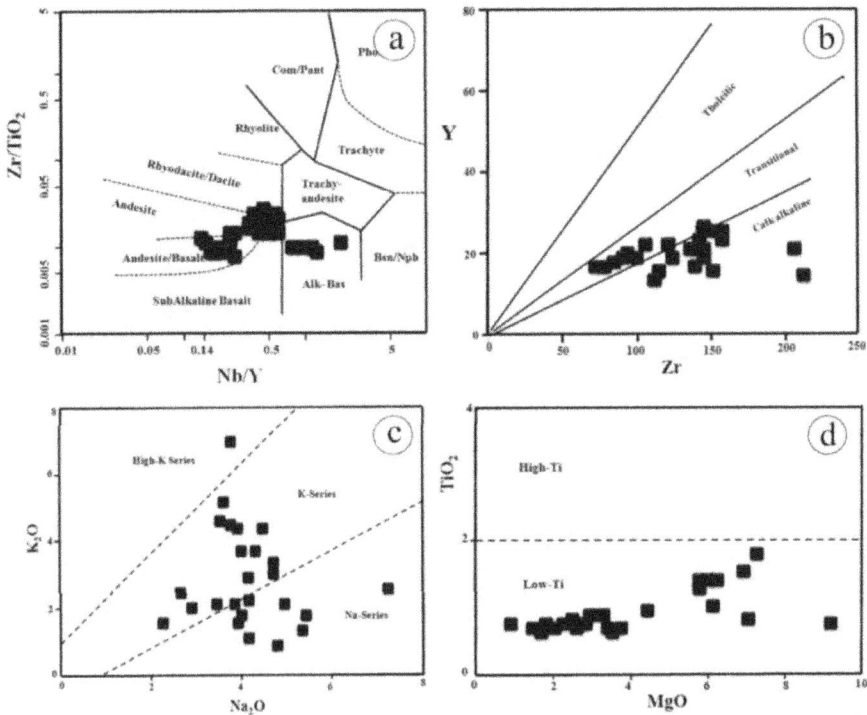

Figure 26.
a) Classification diagrams of volcanic rocks (after [40]). b) Zr vs. Y diagram indicates that the studied area samples belong to transitional to calk alkaline suites. c) K_2O vs. Na_2O diagram (after [41]), showing that the DAEV samples belong to the Na and K-series. d) TiO_2 vs. Mgo diagram (after [42]) indicates that the studied samples have low-Ti contents. (after [15]).

diorite, granodiorite, granite [28], granite, micromonzonite [27], granite intruded by monzodiorite, quartz diorite, monzodiorite, granite, granodiorite [24], quartz monzonite, and quartz monzodiorite [20]. The age determined for these rocks varies from the Paleocene to early Eocene.

Large intrusive masses have the following characteristics: Granite (post-Paleocene), microgranite to granite (late Cretaceous), diorite, monzodiorite (middle-late Cretaceous) in Kuh-e-Baharestan and Kuh-e-Mish (**Figure 30**) [23] which intrude into the late Cretaceous oceanic sediments (**Figure 31**) as well as granite, granodiorite, diorite, and monzonite which are crosscut by microdioritic and diorite-monzonite dykes [53]. Based on lithological, petrological, and geochemical studies on the intrusive rocks of Kuh-e-Mish, it has been demonstrated that they are a mass of granitoid of which a major part is granodiorite and a minor part is tonalite in terms of composition. This granitoid mass is mostly calc-alkaline. It is meta-alumina to slightly per-alumina in terms of the amount of aluminum and it is a HSS (a true hybrid) in terms of origin. The mentioned magma probably originated from the upper mantle and the base of the crust [17]. Goharshahi also believed that the depth of magma production was about 73 km or a little more and its replacement depth was less than 4 km. This intrusive mass was of the granitoid type and was associated both with the volcanic arc granitoid (VAG) of the continental margin and subduction and its continuation after collision (post-collision events). The estimated time

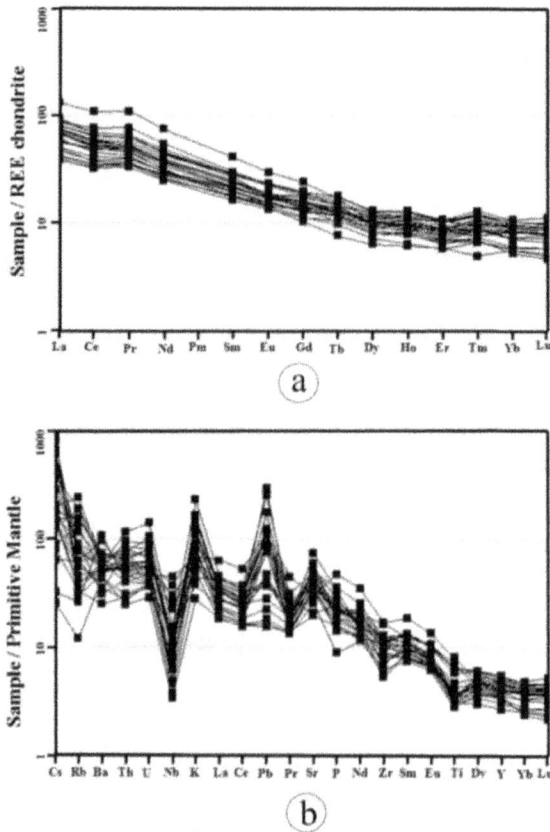

Figure 27.
Chondrite (a) and primitive mantle (b) normalized spider diagrams of DAEV rocks (normalization values are from [36]). (after [15]).

of collision was probably Eocene and the time of granitoid production was about 37 Ma or more (**Figure 32**) [17].

The youngest post-collision volcanic activities: The age of the youngest volcanic activities in the collision zone varies from Miocene to early Quaternary. They are mainly composed of dacite and most of their outcrops are observed in the ophiolites of northern Sabzevar and also behind the volcanic arc in the Joghatay back-arc basin. Dacites are subvolcanic rocks which mostly appear at the intersections of faults. Young dacites crosscut all the older rocks and sediments in the three forms of dome, plug, and dike. From the petrological point of view, they can be classified into the three groups of biotite bearing dacite, amphibole-bearing dacite, and dacite with a small amount of dark minerals [24]. In the western parts of Sabzevar ophiolites, dacitic domes are more abundant both in number and size (**Figure 33**) [24].

In the northernmost part of the collision zone, dacite rocks show an andesitic-dacite composition and appear in the form of domes and lava flows. These lavas cover a large area of the northern margin of the Joghatay back-arc basin. The youngest volcanic cones (Plio-Quaternary in age) can be found in this area ([31]; Radfar [57]) (**Figure 34 and 35**).

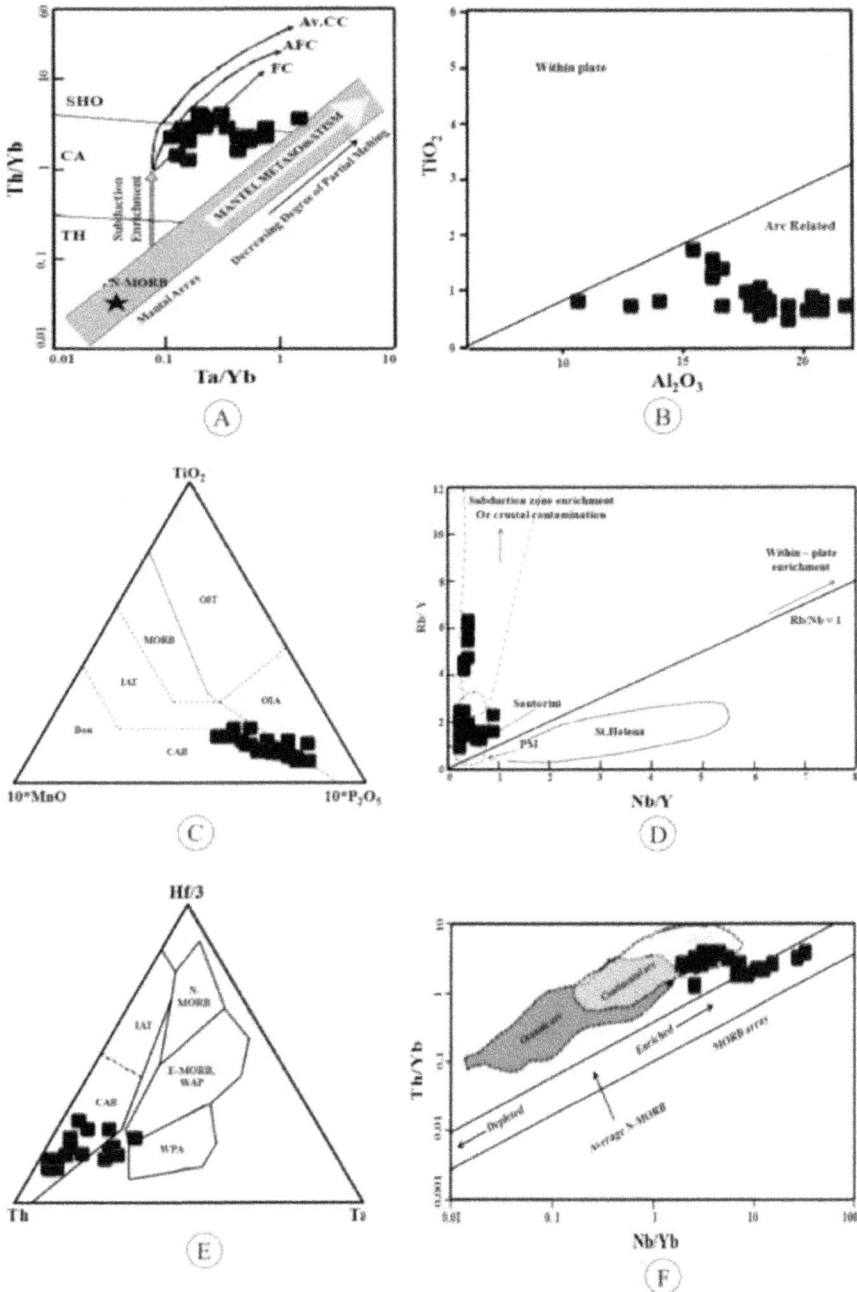

Figure 28.
Tectonic discriminant diagrams for the DAEV rocks. a) Ta/Yb vs. Th/Yb (after [43]). Vectors show inferred effects of fractional crystallization (FC), assimilation- fractional crystallization (AFC), subduction enrichment and mantle metasomatism. b) Al₂O₃ vs. TiO₂ (after [44]). c) TiO₂- MnO-P₂O₅ (after [45]). d) Nb/Y vs. Rb/Y (after [46]). e) Hf-Th-Ta (after [47]). f) Nb/Yb vs. Th/Yb (after [48]). SHO shoshonite, CA calc-alkaline, TH tholeiite, N-MORB normal MORB, E-MORB enrich MORB, OIT Ocean island tholeiitic basalt, LAT island arc tholeiite, BON boninites OIA Ocean island arc basalts, WPA within-plate alkaline. (after [15]).

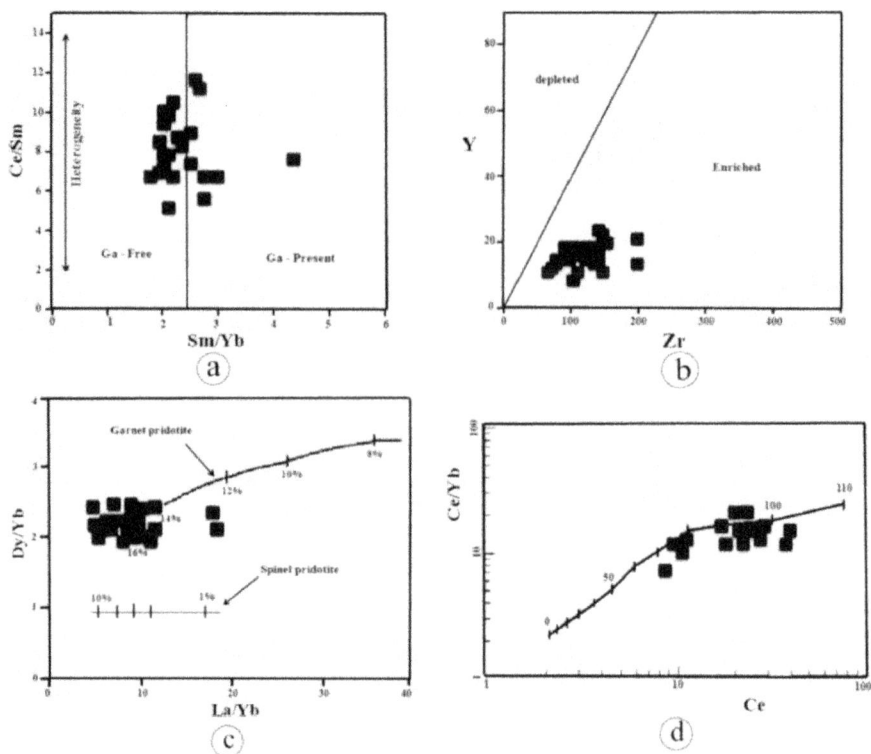

Figure 29.
a) Sm/Yb vs. Ce/Sm plot used to identify the mantel source for DAEV rocks (after [49]). b) Zr vs. Y showing the enriched nature of the DAEV rocks (after [50]). c) La/Yb vs. Dy/Yb plot used to determine the degrees of partial melting of the DAEV source rocks (after [51]). d) Ce vs. Ce/Yb diagram used to determine the depths of the melt segregation for the source for DAEC rocks (after [52]). Ga garnet (after [15]).

Figure 30.
Low elevation of granite masses (post-Paleocene) and microgranite to granite (late Cretaceous) in vicinity of Estaj village (south of Kouh-e-Baharestan).

Figure 31.
Large intrusive masses of diorite, monzodiorite (late Cretaceous to middle Paleocene) which intruded into the late Cretaceous oceanic sediments north of Anjoman Village (south of Sabzevar).

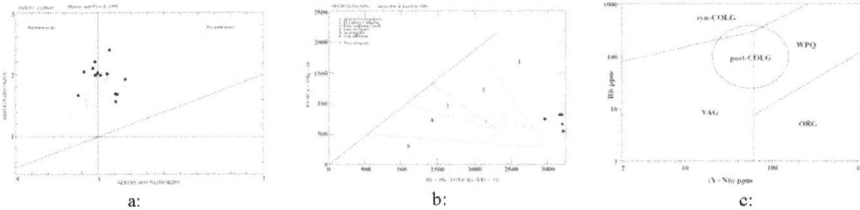

Figure 32.
Geochemistry, chemical classification and location of the post collision granitoids (after [17]) a) the location of granitoid samples from Kuh-e-Mish on the a/CNK vs. ANK Maniar and Piccoli [54] diagram, b) the location of the Kuh-e-Mish and Baharestan granitoids on the Batchelor and Bowden [55] diagram, c) Rb-(Y/+Nb), Nb-Y diagrams for the Kuh-e-Mish granitoid samples on the base of Pearce et al. [56], which separate the locations of the Syn-collision from the within plate volcanic arc and oceanic ridge granites.

Figure 33.
The youngest post- collision volcanic activities in the collision zone (Miocene to early quaternary).

Scale: 1:500,000

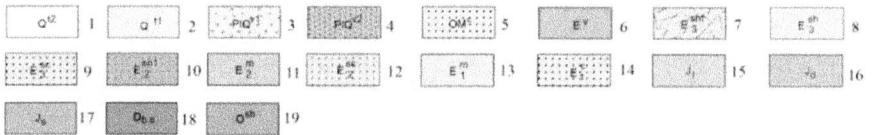

Figure 34.
The youngest volcanic cones (Plio-quaternary in age) and the distribution of lava flows in the northernmost part of the Sabzevar-Nain collision zone (after [31]). Quaternary: 1. Younger alluvium, 2. Older alluvium, Plio- quaternary: 3. Andesitic lava, 4. Dacitic lava and dome, Oligo- Miocene: 5. Conglomerate, Eocene: 6. Arc volcanic rocks, 7. Shale and tuff, 8. Shale, 9. Sandstone and conglomerate, 10. Shale, 11. Marl, 12. Sandstone and conglomerate, 13. Marl, 14. Conglomerate, Jurassic: 15. Limestone, 16. Marly limestone, 17. Shale and sandstone, 18. Limestone and sandstone, 19. Silty shale.

Figure 35.
A Plio- quaternary volcanic cone in the northern most part of the collision zone.

5. Conclusion

Various processes and tectonic events have been influential on the evolution of the Iranian crust. The major events in this regard can be divided into two stages. One of the events in the second stage is the thinning of the crust during the late Jurassic, leading to a period of instability. After this, the Iranian crust broke and then created interior oceanic basins such as Sabzevar-Nain, Nain-Baft, and Sistan-Baluch Basins [2, 4, 5]. The processes of the opening and closure of the Sabzevar-Nain Basin (SNB) began with crustal thinning in the late Jurassic with an east–west trend between the micro-block of central Iran in the south and the Alborz Mountain Ranges in the north. As a deep trough in the middle Cretaceous, the Sabzavar-Nain rift became an oceanic basin at the beginning of the late Cretaceous. At the end of the late Cretaceous, the Sabzevar-Nain Basin reached its final expansion. With the subduction of the oceanic crust under the continental margin of Alborz, the closure of the Sabzevar-Nain Basin began in the early Paleocene. After the collision of the continental margins of central Iran and the Alborz Mountain Ranges, the basin was closed in the middle Eocene. Based on the geometry of the oceanic crust (the direction of subduction is perpendicular to the trench direction), we propose that this continent-continent collision can be classified as a classical collision. Our reasons based on the structural elements of the collision zone are ophiolitic rocks exposed in the suture, remnants of deep oceanic sediments upthrust on both margins, a volcanic arc parallel to the trend of the suture, a back-arc basin, and finally post-collision intrusive masses.

Acknowledgements

We thank Mohammad Reza Mirzaie and Mostafa Khoshduni Farahani for accompanying us in the first field trip. We warmly thank Dr. Fodazi for his petrological advice. We appreciate Mr. Jafarian for providing us with data about the Sheshtamad area and Kuh-e-Bahrestan. Our thanks are also extended to the kind people of the region between Nain and Sabzevar for their hospitality, aids, and information. Finally, we thank Miss Shahosseini for drawing the figures.

Author details

Saidi Abdollah[1*], Khan Nazer Nasser[1], Hadi Pourjamali Zahra[2], Farzad Kiana[2]

1 Geological Survey of Iran, Geological Survey of Iran, Tehran, Iran

2 Rega Zamin Sakht Consulting Engineers, Tehran, Iran

*Address all correspondence to: abdollahsaidi@yahoo.fr

IntechOpen

References

[1] Mc Call GJH. Area Report, East Iran Project, Area No. 1. Iran: Geological Survey of Iran, Tehran; 1985e Report No. 57. 634 PP

[2] Mc Call GJH. Geology of North Makran & South Baluchestan. Geol. Surv. Iran. 1985;57:1-633

[3] Mc Call GJH. A Critique of the Analogy between Archean and Paleozoic Tectonic Based on Regional Mapping of the Mesozoic-Cenozoic Plate Convergent Zone in the Makran, Iran. Liverpool, England: Liverpool University; 2003

[4] Saidi A. Calandrier de la migration Permo-Triassique et morcelement Mésozoïqiue des éléments continentaux de l'Iran: apports de la subsidence et du Paléomagnétique (Thèse de doctorat) Université de Pierre et Marie Curie (Paris 6), Paris, France; 1995

[5] Saidi S, Brunet MF, Ricou LE. Continental accretion of the Iran block to Eurasia as seen from late Paleozoic to early Cretaceous subsidence curves. Geodyn. Acta. 1997;10:189-208

[6] Sadreddini E. Geology and Petrology of the Middle Part of the Sabzevar Ophiolitic Belt [Thesis]. Saarbrücken, Germany: Univ. Saarbrucken; 1974

[7] Alavi Tehrani N. Geology and Petrography in Ophiolite Range NW of Sabzevar (Khorasan), Iran with Special Regard to Metamorphism and Genetic Relation in an Ophiolite Suite [Thesis]. Saarbrücken, Germany: Dissertation der mathematish Naturwissen Schaftlichen Fakultat der Universitat Saarlande; 1976 147P

[8] Dehghani GA. Schwerefeld und krust Naufbau im Iran. Hamburger, Geophys. Einzelschr, Reihe A; Hamburger Universitat. 1981. pp. 54-74

[9] Noghreyan MK. Evolution géochimique, minéralogique et structure d'un édifice ophiolitique singulier, le massif de Sabzevar (partie centrale), NE de l'Iran. The'se e's sci. univ. de Nancy I, France; 1982. pp. 239

[10] Lensch G, Mihn A, Alavi-Tehrani N. Petrography and geology of the ophiolite belt north of Sabzevar/Khorasan (Iran). Neues Jahrbuch für Mineralogie-Abhandlungen. 1977;131:156-178

[11] Lensch G, Mihn A, Alavi-Tehrani N. Major Element Geochemistry of the Ophiolites North of Sabzevar (Iran). Neues Jahrbuch für Geologie und Paläontologie-Monatshefte; 1979. pp. 413-447

[12] Lench G, Mihn A, Alavi-Tehrani N. The Postophiolitic Volcanism North of Sabzevar/Iran. Geology, Petrography and Major Element Geochemistry. Neues Jahrbuch für Geologie und Paläontologie-Monatshefte; 1980. pp. 686-702

[13] Lensch G, Davoudzadeh M. Ophiolites in Iran. Neues Jahrbuch für Geologie und Palaontologie; 1982. pp. 306-320

[14] Baghdadi I. Geochemical, Petrography and Tectonic Setting of the North Sabzevar Eocene Volcanic Rocks. Iran: Geological Survey of Iran; 1999

[15] Ghassemi H, Rezaei-Kahkhaei M. Petrochemistry and tectonic setting of the Davarzan-Abassabad Eocene volcanic (DAEV) rocks, NE Iran. Journal of Mineralogy and Petrology. 2015;109(2):235-252

[16] Khalatbari M, Etessami S. Tectonomagmatic locality of Eocene volcanic rocks of Ahovan region (Semnan). Iranian Geological Quarterly Journal. 2018;12(46):49-64

[17] Goharshahi R. Petrography, geochemistry and tectonic granitoid Pluton near Mish mountain in the south

of Sabzevar. Thesis submitted for degree of Master of Science, for Teacher Education (Kharazmi University), Tehran, Iran; 2001

[18] Saidi A, Akbarpour MR. The Geological Map of Kiasar (South Alborz) 1:100,000. Iran: Geological Survey of Iran; 1983

[19] Saidi A, Vahdati DF. The Geological Map of Sari Quadrangle (North Iran), (1:250,000). Iran: Geological Survey of Iran; 1980

[20] Salamati R, Shafei AR. The Geological Map of Ahmadabad, (1:100,000). Iran: Geological Survey of Iran; 1999

[21] Kolivand H. Kinematic Analysis of Eastern Part of Kharturan Quadrangle (1:250,000) [Thesis]. Iran: Institute for Earth Sciences, Geological Survey of Iran; 2000

[22] Ghaffari NB. Structural Geology Analysis of Middle Cretaceous Syn-Rift Sediments Western Part of Kharturan Quadrangle [Thesis]. Iran: Institute for Earth Sciences, Geological Survey of Iran; 2000

[23] Jafarian MB, Jalali A. The Geological Map of Sheshtamad (1:100,000). Iran: Geological Survey of Iran; 1999

[24] Majidi J, Sahandi MR, Khan Nazer N. The Geological Map of Sabzevar (1:100,000). Iran: Geological Survey of Iran; 2000

[25] IUGS, CGMW. International Chronostratigraphic Chart. International Commission on Stratigraphy; 2015

[26] Lindenberg HG, Gorler K, Ibbekn H. Stratigraphy, Structure and Orogenic Evolution of the Sabzevar Zone in the Area of Oryan (Khorasan), NE Iran. Geodynamic project (Geotraverse) in Iran; 1983 Report No. 51

[27] Akrami MA, Askari A. The Geological Map of Soltanabad,

North-East Sabzevar (1:100,000). Iran: Geological Survey of Iran; 2000

[28] Bahroudi A, Omrani J. The Geological Map of Bashtin, West Sabzevar, (1:100,000). Iran: Geological Survey of Iran; 1999

[29] Eftekhar NJ, Aghanabati A, Hamzehpour B, Baroyant V. The Geological Map of Kashmar Quadrangle, (1:250,000). Iran: Geological Survey of Iran; 1976

[30] Rahmati M, Babakhani AR, Sahandi MR, Khan Nazer N. The Geological Map of Joghatay, North of Sabzevar, (1:100,000). Iran: Geological Survey of Iran; 2000

[31] Tatavousian S, Zohrehbakhsh A, Hamzehpour B, Alavi TN, Sadredini SE, Vaziri Tabar V. The Geological Map of Sabzevar Quadrangle, (1:250,000). Iran: Geological Survey of Iran; 1982

[32] Khalatbari M, Babaei H, Mirzaie M. Geology, Petrology and Tectonomagmatic Evolution of the Plutonic Crustal Rocks of the Sabzevar Ophiolite, NE Iran. Cambridge University, Cambridge University Press; 2013. pp. 1-23

[33] Baroz F, Macaudier J, Montigny R, Noghreyan M, Ohnester M, Rocci G. Ophiolites and Related Formations in the Central Part of Sabzevar Range (Iran) and Possible Geotectonic Reconstructions. The Geodynamic project (Geotraverse) in Iran; 1983 Report No. 51. Geological Survey of Iran

[34] Wood DA, Joron JL, Treuil M. Arc-appraisal of the use of trace elements to classify discriminate between magma series erupted in different tectonic setting. Earth and Planetary Science Letters. 1979;**45**:326-336

[35] Gorton MP, Scandl ES. From continents to island arcs, a geochemical index of tectonic setting for arc-related

within plate felsic to intermediate volcanic rocks, Canadian Mineralogist. 2000;**38**:1065-1073

[36] Sun SS, McDonough WF. Chemical and isotopic systematics of oceanic basalts: Implications for mantle composition and processes. Geological Society of London. 1989;**42**:313-345

[37] Martynov YA, Kimura JI, Khanchuk AI, Rybin AV, Chashchin AA. Magmatic sources of quaternary lavas in the Kuril island arc: New data on Sr and Nd isotopy. Doklady Earth Sciences. 2007;**417**(8):1206-1211

[38] Tian L, Castrillo PR, Hawkins JW, Hilton DR, Hanan BH, Pietruszka AJ. Major and trace element and Sr-Nd isotope signatures of lava from central Lau Basin: Implications for the nature and influence of subduction components in the back arc mantle. Journal of Volcanology and Geothermal Research. 2008;**178**:657-670

[39] Shahosseini A, Ghassemi H. Petrology and petrography investigation on the intrusive masses in the Noukeh area, north-East Semnan. In: Presented in the Second International Geological Congresses. Iran: Geological Survey of Iran; 2007

[40] Winchester JA, Floyd PA. Geochemical discrimination of different magma series and their differentiation products using immobile element, geology. Chemical Geology. 1977;**20**:249-287

[41] Middlemost EAK. The basalt clan, earth. Sci. Rev. 1975;**11**:337-364

[42] Saadat S, Stern CR. Petrochemistry and genesis of olivine basalt from small monogenetic parasitic cones of Bazman stratovolcano, Makran arc, southeastern Iran. Lithos. 2011;**125**:607-619

[43] Pearce JA. Role of the sub-continental lithosphere in magma genesis at active margins. In: Hawkesworth CJ, Norry MJ, editors. Continental Basalts and Mantle Xenolite. Cheshire: Shiva; 1983. pp. 230-249

[44] Muller D, Rock NMS, Grove DI. Geochemical discrimination between shoshonitic and potassic rocks from different tectonic setting; a pilot study. Mineralogy and Petrology. 1992;**46**:259-289

[45] Mullen ED. Mno/Tio2/P2o5: A minor element discriminant for basaltic rocks of oceanic environment and its implications for petrogenesis. Earth and Planetary Science Letters. 1983;**62**:53-62

[46] Esporanca S, Crisci M, de Rosa R, Mazzuli R. The role of the crust in the magmatic evolution of the island Lipari (Aeolian Islands, Italy). Contributions to Mineralogy and Petrology. 1992;**112**:450-462

[47] Wood DA. The application of a Th-Hf-Ta diagram to problems of tectonomagmatic classification and to establishing the nature of crustal contamination of basaltic lava of the British tertiary volcanic province. Earth and Planetary Science Letters. 1980;**50**:11-30

[48] Pearce JA, Peate DW. Tectonic implications of the composition of volcanic arc magmas. Annual Review of Earth and Planetary Sciences. 1995;**23**:251-285

[49] Coban H. Basalt magma genesis and fractionation in collision and extension related provinces: A comparison between eastern, central and western Anatolia. Earth-Science Reviews. 2007;**80**:219-238

[50] Abu-Hamatteh ZS. Geochemistry and petrogenesis of mafic magmatic rocks of the Jharo Belt, India: Geodynamic implication. Journal of Asian Earth Sciences. 2005;**25**:557-581

[51] Thirlwall MF, Smith TE, Graham AM, Theodorou N, Hollings P, Davidson JP, et al. Resolution of the effects of crustal assimilation, sediment subduction, and fluid transport in island arc magma: Pb-Sr-Nd- O isotope geochemistry of Grenada, Lesser Antilles. Geochem. Cosmochim. Acta. 1996;**60**:4785-4810

[52] Ellam RM. Lithospheric thickness as a control on basalt geochemistry. Geology. 1992;**20**:153-156

[53] Naderi MN, Shojai KH, Bahremand M. The Geological Map of Shamkan, South-East of Sabzevar (1:100,000). Iran: Geological Survey of Iran; 2000

[54] Maniar PD, Piccoli PM. Tectonic discrimination of granitoids. GSA Bulletin. 1989;**101**:635-643

[55] Batchelor RA, Bowden P. Petrographic interpretation of granitoid rock series using multicationic parameters. Chem., Geol. 1985;**48**:43-55

[56] Pearce JA, Harris NB, Wand Tindle AG. Trace element discrimination diagrams for the tectonic interpretation of granitic rocks. Jour. Petrol. 1989;**25**(part 4):956-983

[57] Radfar J. Geological Map of Safiabad (1:100,000). Iran: Geological Survey of Iran; 1999

Time-Series Analysis of Crustal Deformation on Longstanding Transcurrent Fault: Structural Diversity along Median Tectonic Line, Southwest Japan, and Tectonic Implications

Yasuto Itoh

Abstract

The Median Tectonic Line (MTL) along the longstanding convergent margin of eastern Eurasia has been activated intermittently since ca. 100 Ma. In its incipient phase, propagating strike slips on the MTL generated an elongate pull-apart depression buried by voluminous clastics of the Late Cretaceous Izumi Group. In this study, the complicated deformation processes around this regional arc-bisecting fault are unraveled through a series of quantitative analyses. Our geological survey of the Izumi Group was exclusively conducted in an area of diverse fault morphology, such as jogs and steps. The phase stripping method was introduced to elucidate the time sequence of cumulative tectonic events. After stripping away the initial structure related to basin formation, neotectonic signatures were successfully categorized into discrete clusters originating from progressive wrenching near the active MTL fault system, which has been reactivated by the Quaternary oblique subduction of the Philippine Sea Plate. The method presented here is simple and effective for the detection and evaluation of active crustal failures in mobile belts where records of multiphase architectural buildup coexist.

Keywords: oblique subduction, transcurrent fault, pull-apart basin, neotectonics, Median Tectonic Line (MTL)

1. Introduction

Island arcs exhumed along convergent plate boundaries surrounding the globe are under intensive stress, and hence, they are sites where diverse tectonic forms can be found. Among such features, arc-bisecting faults activated under a regime of oblique subduction [1] have great importance with regard to regional deformation and terrane rearrangement within mobile belts.

The Japanese Islands exhibit the widest range of active landforms since their recent tectonic processes are under the control of four interacting plates around the east Eurasian margin (see **Figure 1** inset). This study focuses on the south-western part of the arc-trench system, in which transient convergent modes of the Philippine Sea Plate along the Nankai Trough (**Figure 1**) have governed its architectural development.

Except for Kyushu Island in front of a chain of underthrusting bumps on the oceanic plate (Kyushu-Palau Ridge in **Figure 1**), ongoing neotectonic events within southwest Japan can simply be understood by evaluating the obliqueness of plate convergence and related activity levels on a regional arc-bisecting fault, the Median Tectonic Line (MTL).

Here, the research focus is placed on the northwestern part of Shikoku Island (**Figure 1**) because a previous study [3] identified deformation events along with the MTL based on a detailed geologic survey of a Quaternary unit. In the following sections, the tectonics of southwest Japan since the Cretaceous are reviewed, and original results are presented based on an extensive structural analysis making use of the phase stripping method to distinguish recursively overlaid tectonic events, and active crustal damage zones along the MTL are then extracted. This study provides a precedent for further quantitative geologic explorations of mobile belts, where superimposed records of the Earth's evolution remain untouched.

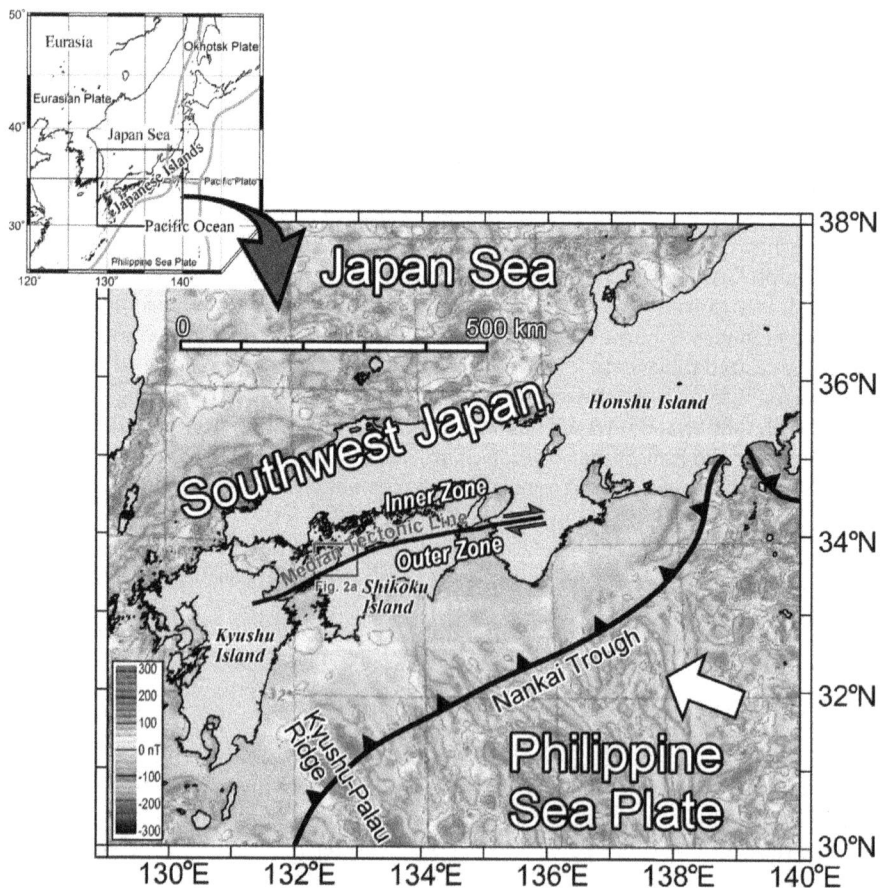

Figure 1.
Index maps for the east Eurasian margin and southwest Japan. Offshore geomagnetic anomalies are after [2]. The open arrow shows the present relative motion of the Philippine Sea Plate.

2. Geological background

2.1 Longstanding activities on the MTL

Most active faults in the Japanese Islands have been vitalized under the complex and unstable regime of plate tectonics through the Quaternary (inset in **Figure 1**). The MTL, however, has an exceptionally long history of activity that dates back to the Cretaceous, when the vigorous northward motion of the Izanagi Plate [4] triggered a breakup of the continental rim and sinistral slips on the proto-MTL. Note that southwest Japan in that period still constituted a part of the east Eurasian margin because the Japan Sea (**Figure 1**) is a back-arc basin that began to open in the mid-Cenozoic [5], although its spreading center is not obviously identified (see the geomagnetic anomalies in **Figure 1** after [2]).

After a period of substantial dormancy during the Paleogene era, the MTL was reactivated under control of the intermittent convergence of the Philippine Sea Plate that had developed during the late Cenozoic around the northwestern Pacific [6]. Southwest Japan at the beginning of the Pliocene was a site of strong inversion. N-S compressive structures that simultaneously built up along the Japan Sea back-arc shelf are suggestive of resumed northward movement of the oceanic plate [7]. Reflecting such a tectonic context, the MTL in the Pliocene stage acted as a low-angle fault, on which watershed mountainous ranges successively emerged [8].

A close observation of the geologic characteristics along the subduction zone [9] found that the Philippine Sea Plate changed its converging azimuth to counterclockwise at approximately 2–1 Ma. This significant fact implies that the west-northwestward motion of the plate around southwest Japan (arrow in **Figure 1**) inevitably enhanced right-lateral activity on the regional fault. Cumulative lateral separation on the present MTL (e.g., [10]) matches well with this tectonic model. To make an explicit distinction between this and older MTL activity, the Quaternary fault driven by dominantly dextral shear is hereafter referred to as the MTL active fault system (MTLAFS).

2.2 Izumi Group filling a regional pull-apart basin

As stated above, the MTL was activated as a left-lateral fault along the Eurasian margin during the Late Cretaceous. Around a propagating termination of the regional rupture, enormous trench-parallel pull-apart basins developed and were promptly buried by voluminous marine siliciclastics that are collectively named the Izumi Group [11]. This group is exposed in an area 300 km long by 10–20 km wide along with the MTL, and it topographically coincides with the core of watershed ranges uplifted during the Pliocene.

Ruled by the general development processes of pull-apart basins [12] and contraction in the Pliocene event of N-S inversion, the stereotypical geologic architecture of this Cretaceous unit is an east-plunging syncline. Such a monotonous structural feature is crucial for isolating neotectonic deformation trends by means of a filtering method that is fully explained in Section 3.

2.3 Gunchu Formation: An indicator of active basin formation and exhumation

2.3.1 Geologic system in the coastal area on the Iyonada Sea

Figure 2 depicts the geologic system around the study area in the northwestern part of Shikoku Island [13, 14]. In a broad way, the Cretaceous basin fill of the Izumi Group is in contact with the high-pressure Sanbagawa metamorphic

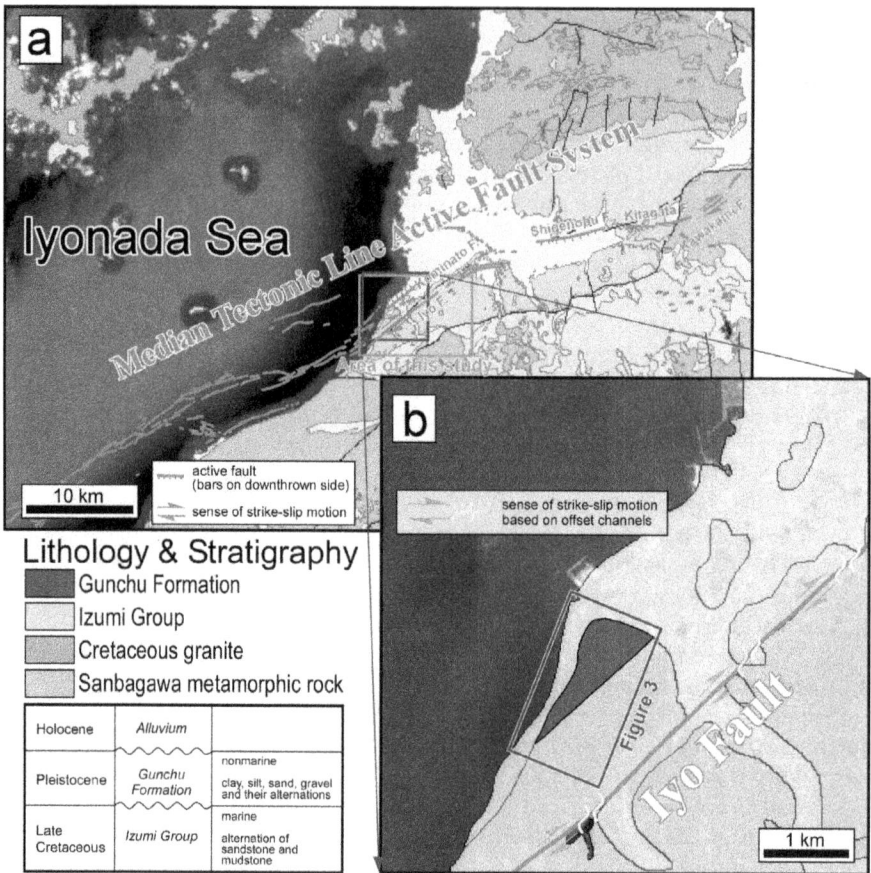

Figure 2.
Geological summary of the northwestern part of Shikoku Island. The geologic system is after [13]. (a) Base map and morphology of the Median Tectonic Line active fault system are after [14–16], respectively. See **Figure 1** *for the mapped area. (b) Sense of active movements on the Iyo Fault is after [17].*

rock to the south, which ascended rapidly in a subduction zone during the sinistral phase of the MTL (cf. Section 2.1). In contrast, the Izumi Group rests on Cretaceous granitic rock accompanied by low-pressure Ryoke metamorphic rock on its northern border. It is noted that a monoclinal nonmarine Pleistocene formation known as the Gunchu Formation is cropped out on the coast of the inland sea. Its steep structure is a result of Quaternary activity on the MTLAFS that runs through northwest Shikoku and the southern coastal zone of the Iyonada Sea [14–17].

2.3.2 Stratigraphy and structural trend

The fluvial sequence of the Gunchu Formation was originally divided into three members [18, 19] and redivided by Itoh [3] into two from the viewpoint of the provenance of pre-Neogene clasts. As delineated in **Figure 3**, the Gunchu Formation lies unconformably on an eroded surface of the Izumi Group and has a steep homoclinal structure. Offsets of its stratigraphic boundaries point to the presence of faults crosscutting the unconsolidated strata. Kitabayashi et al. [20] dated zircon grains separated from volcanic ash intercalated in the basal part of the lower member to be 2.2 ± 0.3 Ma (FT age) and 2.13 ± 0.05 Ma (U–Pb age), whereas an ash layer near the top of the same member yielded zircon ages of 1.8 ± 0.2 Ma and 1.92 ± 0.05 Ma based on FT and U–Pb methods, respectively [3].

2.3.3 Sedimentology and basin-forming process

The Gunchu Formation is a stack of channel and bar deposits with minor facies fluctuation that reflects the migration of channels within an alluvial basin. Apart from such phenomena, Itoh [3] recognized drastic changes in gravel compositions probably linked to successive exhumations of hinterlands driven by tectonic uplift.

Figure 4 exemplifies the compositional variety observed in the fault-bounded Block 1 (**Figure 3**) of the Gunchu Formation. As depicted in the pie charts, the lower member contains a considerable amount of granitic pebbles, the U–Pb ages of which range from 106.5 to 93.8 Ma, as measured for five clusters in four sites [3]. Although these radiometric dates reflect the Cretaceous plutons extensively found in southwest Japan, igneous rocks in that period are enigmatically absent around the present study area. The base of the upper member is defined by an abrupt influx of high-pressure metamorphic clasts. The concentration of schist-originated material is so extreme that some outcrops possess a blue-greenish appearance. The content of metamorphosed

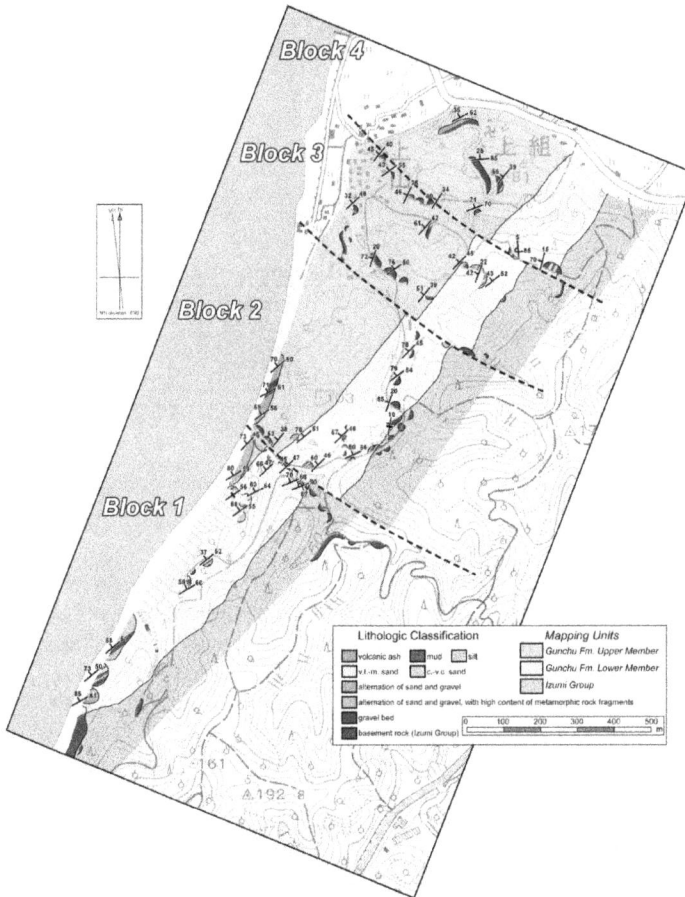

Figure 3.
Geologic map of the Pleistocene Gunchu Formation after Itoh [3]. See **Figure 2b** *for the mapped area. Background topography is a part of the "Kaminada" locality map at 1:25,000, published by the Geospatial Information Authority of Japan.*

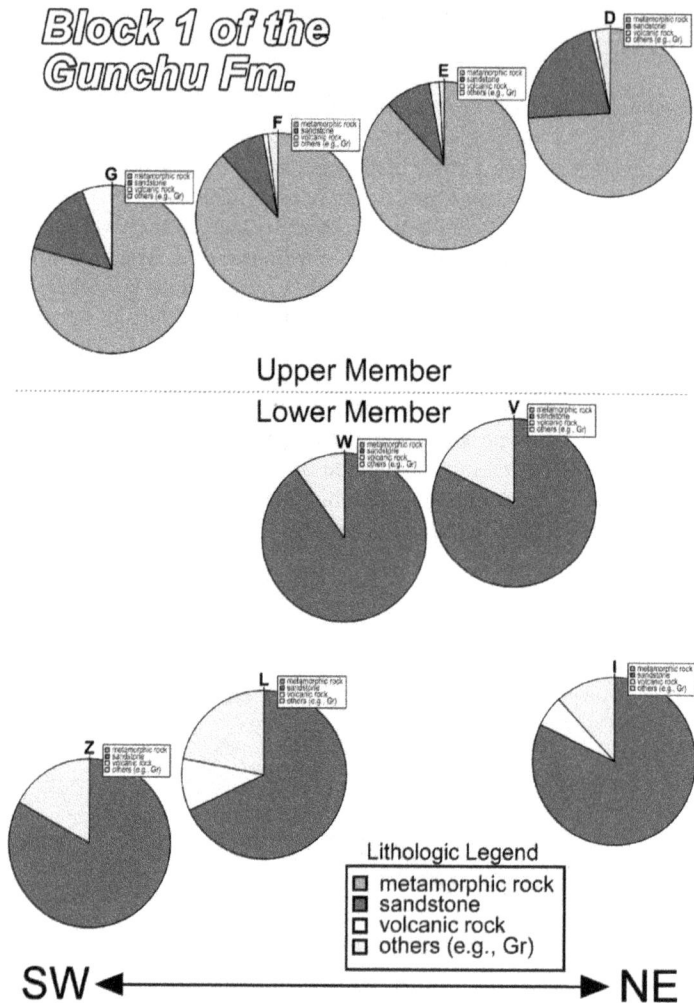

Figure 4.
Spatiotemporal variation in gravel compositions for Block 1 of the Gunchu Formation after [3]. Relative abundance of rock facies was determined from 100 data points for each observation station.

gravels tends to decrease upward, and sandstones derived from the nearby Izumi Group become prevalent in the uppermost part of the sedimentary unit.

Itoh [3] attempted to reconstruct the Quaternary basin-forming process based on the above-stated spatiotemporal changes in gravel composition, and additionally, paleocurrent directions that were determined from the imbricated structure of clasts for the same 26 sites in the compositional analysis. **Figure 5** schematically shows the neotectonic evolution of the northwestern part of Shikoku Island. As described previously, granitic pebbles ubiquitously detected in the lower part of the Pleistocene sediments yielded numerical ages concordant with those of the Cretaceous intrusions within the inner zone (north of the MTL; see **Figure 1**). Significant scatter in the U–Pb ages implies that the granites were not derived from a local intrusive body having a uniform cooling history but from asynchronously emplaced plutons, which are widely exposed on the south side of Honshu Island (**Figure 1**). Thus, all the available data point to an assumption that the depocenter at an early stage was located around the area now occupied by the Gunchu Formation (**Figure 5a**).

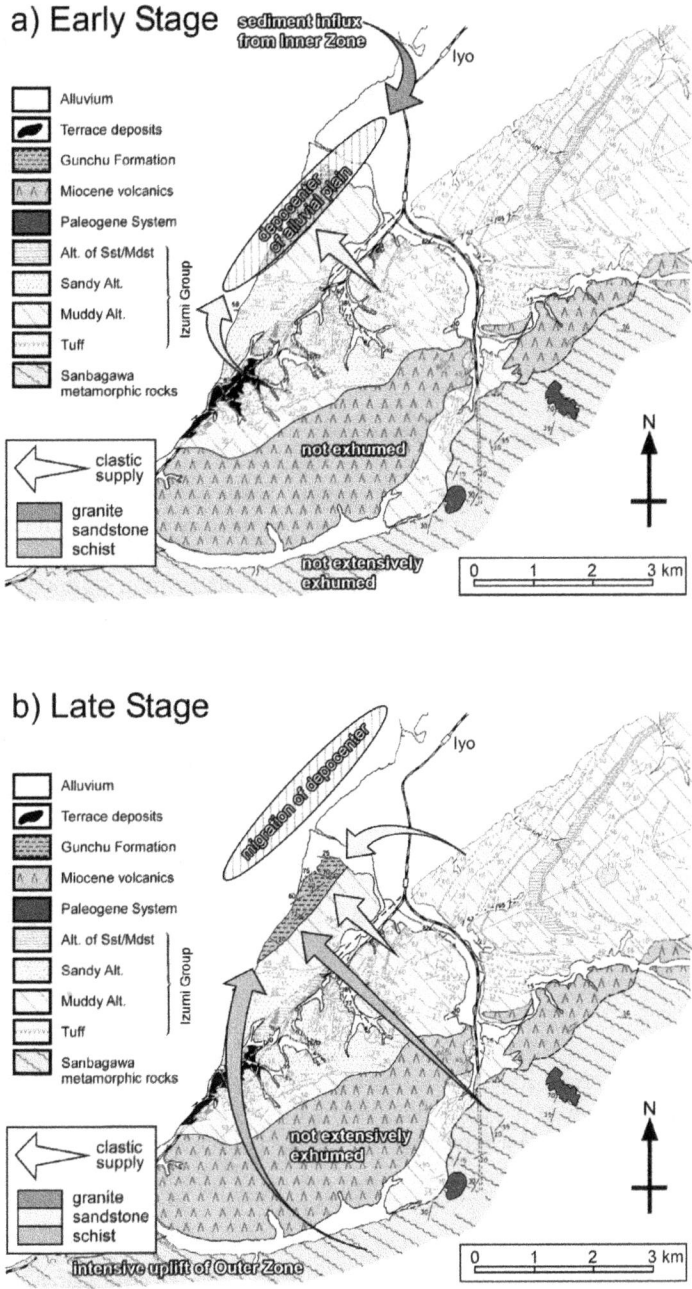

Figure 5.
Paleoenvironments during depositional stages of the Gunchu Formation after [3]. Base geologic map is after [13].

As for the setting of later stages, the massive influx of the Sanbagawa metamorphic rock demonstrates regional uplift in the outer zone (south of the MTL; see **Figure 1**). It is, however, intriguing that the upper member contains a few of the Miocene volcanics that extensively cover the area between the Gunchu basin and the Sanbagawa terrane. Itoh [3] thereby assumed that the region of intensive uplift and erosion progressively expanded northward during the mid-Pleistocene. This resulted in a sharp increase of upward sandstone clasts coming from the nearby Izumi Group and brought about seaward migration of the depocenter, in which clastics derived from the inner zone are trapped. This is the reason why granite pebbles disappeared from the upper member. This tectono-sedimentary model is summarized in **Figure 5b**. The regional and incremental contraction eventually urged the recent basin fills to build up a steep monoclinal structure along with the coastline.

Our review has thus shown a longstanding history of MTL activity and relevant processes of basin formation and deformation. In the following sections, more extensive and quantitative analyses of fault-related tectonics are discussed based on the results of this study.

3. Analysis

3.1 Application of phase stripping method

Previous studies have shown that the structural and sedimentological features of the Pleistocene Gunchu Formation have preserved neotectonic information linked to activity on the MTLAFS. As for advanced research on the fault-bound tectonics, however, confined exposure of the fluvial unit hinders more regional analysis. Therefore, this study focuses on the Cretaceous Izumi Group, which is distributed along the proto-MTL. Although the widespread sediments probably record recent episodes, they also reflect the initial architecture built up during the growth of pull-apart basins. In conventional geological mapping, neotectonic signatures related to active deformation are interpreted as secondary features of basin-forming processes or are excluded as noisy data near local faults. Thus, an overlapped event usually ends up as misread or dead information in a one-sided interpretation.

During the course of research on the easternmost part of the MTLAFS, Itoh and Iwata [21] developed a simple method to separate the multiphase deformation. Their "phase stripping method" regards the geologic architecture revealed through field surveys as an integration of an initial tilting and a younger event. The total structure directly measured on an outcrop, C, is expressed using a matrix product as follows:

$$C = BA \qquad (1)$$

where A is the trend acquired during the initial tectonic phase, a pull-apart basin evolution in this case, and B is the pursued phase formed in a recent period. At each observation point, B can be determined using the known structural data as follows:

$$B = BAA^{-1} = CA^{-1} \qquad (2)$$

In this study, the author applied this simple calculation to field data obtained from 720 outcrops of the Izumi Group.

Figure 6 represents the concrete procedure of phase stripping. In spite of later disturbance, a general structural trend can be identified based on 280 and 440 points of field data in the northeastern and southwestern blocks of the three-year geologic survey, respectively.

As mentioned before, the typical architecture of the Izumi Group, A in Eq. (1), is a monotonous east-plunging syncline developed through the progradational burial of a series of pull-apart basins and is delineated by red contours in the figure. For reference, the Pleistocene Gunchu Formation is located near the north corner of the southwestern block.

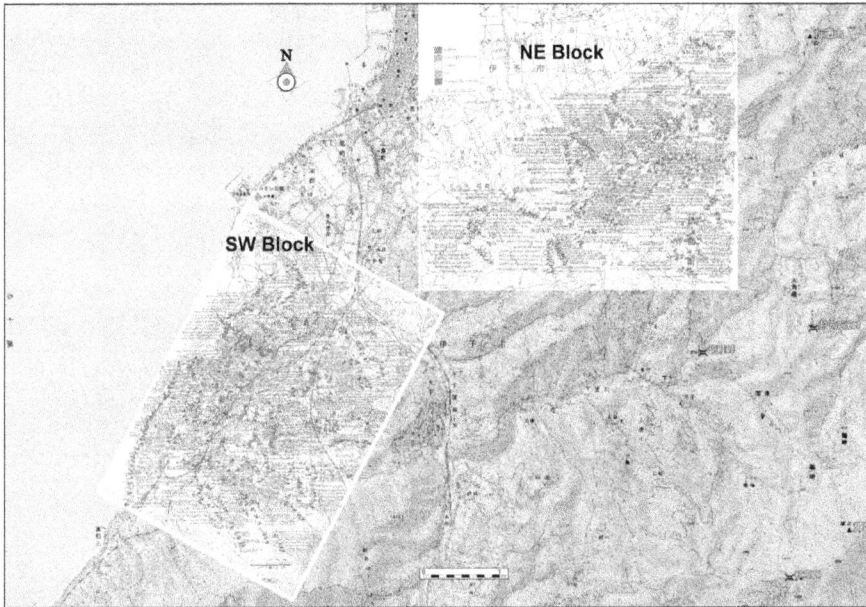

Figure 6.
Base map for the phase stripping analysis of the Izumi Group. See magenta enclosure in **Figure 2a** *for the mapped area. Detailed topographies of the two analyzed blocks are parts of the "Kaminada" and "Gunchu" localities at 1:25,000, published by the Geospatial Information Authority of Japan. There are 280 and 440 data points in the northeastern and southwestern blocks, respectively. Thin red lines and attached dip angles represent the initial east-plunging synclinal structure of the Cretaceous basin fill.*

3.2 Detection of damage zones

The analytical results of this study are summarized in **Figure 7**. As shown by magenta lines, a previous geomorphological study [17] recognized two branches of the MTLAFS, namely, the minor Kominato Fault and the major Iyo Fault, for which right-lateral offsets were confirmed (**Figure 2b**).

Elaborate field observations led the author to realize that the density of visible fractures in the Izumi Group varies considerably under the possible influence of local deformation. To perform a quantitative evaluation, the fracture density was measured at 18 sites along with the coast. At every station, all the visible cracks residing within a massive sandstone bed were counted on a 10 × 10 grid of graticules ruled in 3-cm divisions, which were drawn on a transparent polyethylene film pasted on the outcrop surface. All the fracture counts at 18 stations are shown in **Figure 7**, and histograms in the figure present selected results of such measurements.

In the same figure, regions of severe deformation are highlighted by red symbols for steep and overturned bedding attitudes of phase B in Eq. (2). As for the northeastern part of the analyzed area, a major damage zone lies on or near the Iyo Fault, and minor trends extend along gorges subparallel to the main active fault. In contrast, more diverse structural disturbances characterize the southwestern area. Some data indicating intense motions in phase B are aligned near the straight scarp of the Iyo Fault, but others show elongated clusters with different orientations, implying the existence of unknown active failure zones. In both the analyzed areas of **Figure 7**, remarkable phase B deformations are found irrelevant to the initial synclinal structure of phase A. In the next section, the recent progressive deformation process is examined by integrating geological evidence, and a neotectonic model of the MTLAFS is derived that matches the structural framework.

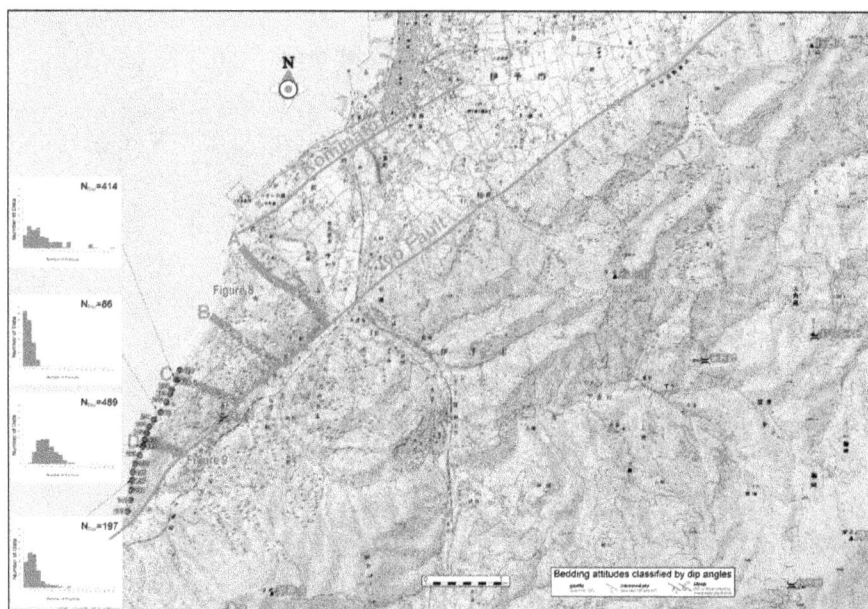

Figure 7.
Extracted neotectonic features based on the phase stripping method. The mapped area is the same as in **Figure 6.**
Previously reported active faults (magenta lines) are after [17]. Refer to the main text for trends A to D (thick gray lines), fracture numbers at 18 stations pasted along the coast, and histograms of the fracture density. Bedding attitudes developed under recent stress are categorized as gentle (yellow, less than 30°), intermediate (orange, between 30° and 60°), and steep (red, 60° or more, including overturned structures).

4. Discussion

4.1 Subordinate active faults

The phase stripping method applied to the Izumi Group distributed around northwestern Shikoku Island successfully delineated recent damage zones in connection with activities on the MTLAFS. As for the wedge between the coastline and the Iyo Fault, four-fault traces were identified based on linear clusters of the strong phase *B* motions (shown by red symbols in **Figure 7**) and are depicted as gray zones A to D in the figure. Trends A and B are concordant with faults cutting the Pleistocene Gunchu Formation (see **Figure 3**). Trends C and D are recognized as deformation zones that branch off from the Iyo Fault and appear to be related with coastal stations showing extreme concentrations (414 and 489 counts for each measurement) of fractures (histograms in **Figure 7**). Such parallel trends aside, the phase *B* data tend to suggest stronger deformation along with the Iyonada Sea coast. Although numerous offshore faults have been found through previous studies (**Figure 2a**), the detailed structure of the shoreline remains ambiguous because research vessels cannot perform sounding surveys in very shallow water. The present results imply an offshore stepped extension of the Iyo Fault that provokes activity on the subordinate faults.

4.2 Actual deformation processes recorded in outcrops

As mentioned above, the Pleistocene Gunchu Formation possesses a steep homoclinal structure suggestive of strong neotectonic deformation. **Figure 8** represents the unconformity between the Izumi Group and the Gunchu Formation

North　　　　　　**Izumi Group**　　　　　　**South**
(massive m.-f. sandstone)

Gunchu Fm.
(unconsolidated
gravel bed)

intercalating flaky mudstone

undulated boundary

unconformable interface is buried by
angular f. sandstone cobbles with
interstitial mudstone small pebbles

matrix-supported,
steep bedding

clast-supported,
overturned bedding

0　　　　5　　　　10 m

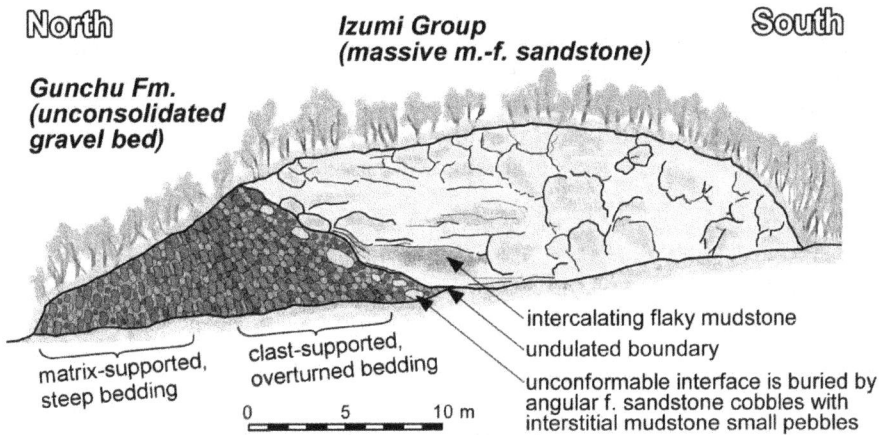

Figure 8.
Sketch of the unconformable relationship between the Cretaceous Izumi Group and the Pleistocene Gunchu Formation after [3]. See **Figure 7** *for the outcrop location. Bedding attitudes for the Pleistocene unit imply syn-sedimentary structural growth.*

first described by Itoh [3]. Note that the basal part of the Pleistocene gravel strata exhibits overturned bedding, which graduates upward into steep but normal bedding attitudes. Such a change indicates syn-depositional structural buildup. Another point is that the phase *B* data near this exposure (see **Figure 7** for its location) are in the gentle level of recent motions (tilting shallower than 30°). These geologic lines of evidence seem to suggest that a flexure zone developed under a compressive regime during the uplift of the hinterlands, roughly coinciding with the recent unconformable interface.

A sketch of important outcrops is presented in **Figure 9** (see **Figure 7** for the location). Steep tilting in its western part is likely to be affected by lateral motions on the adjacent Iyo Fault. It is notable that the Izumi Group in the central and

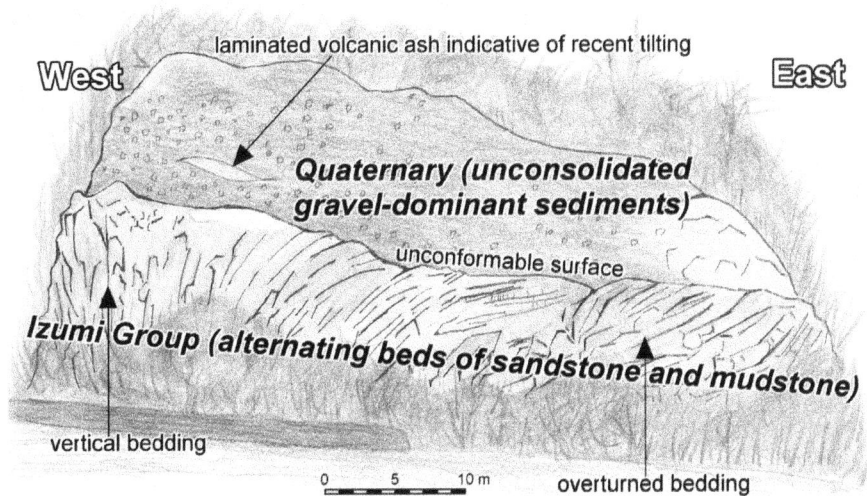

laminated volcanic ash indicative of recent tilting

West　　　　　　　　　　　　　　　　　　　　**East**

Quaternary (unconsolidated
gravel-dominant sediments)

unconformable surface

Izumi Group (alternating beds of sandstone and mudstone)

vertical bedding

0　　5　　10 m

overturned bedding

Figure 9.
Sketch of an outcrop in which accumulation processes of neotectonic deformation are observed. See **Figure 7** *for the outcrop location. Refer to the main text for the structural interpretation.*

eastern parts of the outcrop shows overturned attitudes. Namely, as we move away from the active fault, the deformation of the strata seems to become stronger. This paradoxical phenomenon implies that an unreported zone of active contraction extends on the eastern side because the intensive signatures of phase *B* are confirmed within broadband around this observation point (see **Figure 7**).

An unconformable surface of the Izumi Group in this exposure is overlain by gravel-rich unconsolidated sediments. This unit seems to be gently inclined eastward, a tendency endorsed by the sedimentary structure of an intercalation of fine volcanic ash (**Figure 9**). Neotectonic activities on the MTLAFS thus resulted in severe deformation and exhumation of the Cretaceous basin fill, and the overlying fluvial deposits are significantly tilted under lingering tectonic stress.

4.3 Integrated neotectonic model

Figure 10 depicts an integrated model of neotectonic processes on a part of the MTLAFS, which is modified from the prototype submitted by Itoh [3]. First, an unknown transcurrent fault having a right offset from the Iyo Fault is postulated on the basis of active structures in the Iyonada Sea (**Figure 2**). Such a stepped

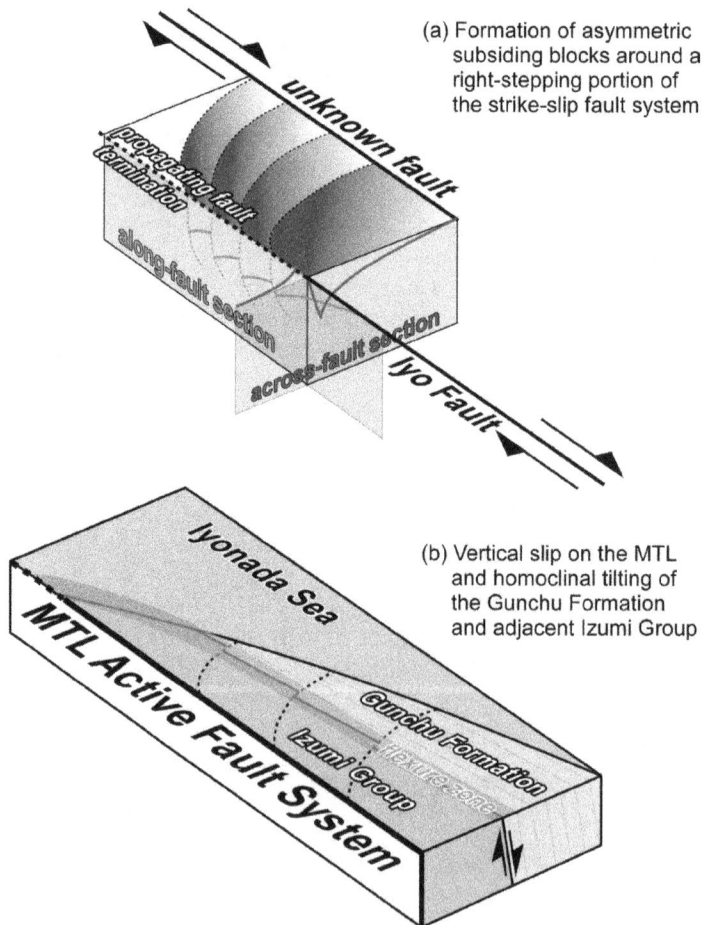

Figure 10.
Schematic diagrams delineating the probable evolutionary history of the study area adjacent to an active transcurrent fault system. It is modified from Itoh [3] based on the present results.

morphology accompanied by propagation of fault termination enhanced activity on secondary faults bridging the primary strike-slip features [22]. Compartments divided by the subordinate faults were systematically tilted and rapidly buried by onlapping sediments (**Figure 10a**). Next, a rising contractional regime brought about progressive uplift of the hinterlands and seaward migration of the Pleistocene pull-apart sag. Strong tectonic stress eventually provoked steep tilting of the Gunchu Formation on a narrow flexure zone (see **Figure 10b**) and confined the deformation of the Izumi Group. The latest dextral motions on the MTLAFS may have induced reactivation of the crosscutting faults, which is inferred from the trends of recent intensive deformation (**Figure 7**).

This study has demonstrated that a long and complicated motion history of the MTL governs the architectural development of nearby geologic terranes. As it is a crustal break under the control of subduction modes of oceanic plates, the fault-related tectonics may have a wider influence over evolutionary processes for the island arc. For example, Itoh et al. [23] performed a volumetric analysis of the Iyonada Sea based on gravity anomalies and found a gigantic buried basin resting against a 4-km-deep scarp of the MTLAFS. Thus, a regional tectonic model of active margins should be built using multidisciplinary research to shed light on the deep interior of the earth.

5. Conclusions

1. Vigorous tectonic development of the southwestern Japan arc is illuminated in light of the motion history of an arc-bisecting fault, the MTL, which has been activated intermittently under transient subduction of oceanic plates since ca. 100 Ma. In its early stage, sinistral motions on the propagating fault produced an elongated pull-apart depression that was buried by the Cretaceous marine sediments of the Izumi Group. In contrast, the latest stage is characterized by dextral movements provoked by oblique convergence of the Philippine Sea Plate, which urged fault-parallel basin formation and massive uplift and erosion of the hinterlands.

2. The phase stripping method was applied to 720 points of structural data for the Izumi Group distributed along the MTL running through northwestern Shikoku Island. After removing an initial trend (east-plunging syncline), the present analysis succeeded in the isolation of a large amount of data suggesting considerable younger deformations, some of which are aligned parallel to the MTLAFS. Others, however, constitute linear trends in different directions, indicating unknown faults. Prominent trends near the coast of the Iyonada Sea are connected with faults cutting across the Pleistocene Gunchu Formation and are, therefore, regarded as active features.

3. Actual processes of structural growth were reconstructed based on outcrop observations. Geologic structures of the Izumi Group exposed within a zone of severe recent deformation imply the presence of an unreported reverse fault. Bedding attitudes of the Gunchu Formation are suggestive of syn-sedimentary tectonic disturbance. In this study, thus, the MTLAFS tends to be accompanied by recurrent contractional features, which may have been built up during episodic increases in compressive stress, reflecting fluctuations in convergent modes of the oceanic plates.

4. An integrated neotectonic model of architectural evolution in the study area was developed by assuming a right-stepping portion on the MTLAFS. Propa-

gation of dextral fault termination at such a structural singularity inevitably formed an active pull-apart basin divided by subordinate faults, which were promptly buried by the Quaternary clastics derived from uplifted hinterlands. A succeeding compressive regime triggered the seaward basin migration and eventual rollover of the recent basin fill. The latest dextral movements on the MTLAFS may have resulted in activity resuming on the basin-dividing faults.

Acknowledgements

The author is grateful to Shigekazu Kusumoto for his thought-provoking discussion during the course of this study. Some of the graphic art was prepared by Rin Itoh.

Author details

Yasuto Itoh
Osaka Prefecture University, Osaka, Japan

*Address all correspondence to: itoh@p.s.osakafu-u.ac.jp

IntechOpen

References

[1] Fitch TJ. Plate convergence, transcurrent faults, and internal deformation adjacent to southeast Asia and the western Pacific. Journal of Geophysical Research. 1972;77: 4432-4460

[2] Geological Survey of Japan, AIST, CCOP (Coordinating Committee for Coastal and Offshore Geoscience Programmes in East and Southeast Asia), editor. Magnetic Anomaly Map of East Asia 1:4,000,000 CD-ROM Version (2nd Edition), Digital Geoscience Map P-3. Tsukuba: Geological Survey of Japan, National Institute of Advanced Industrial Science and Technology; 2002

[3] Itoh Y. Gunchu Formation—An Indicator of Active Tectonics on an Oblique Convergent Margin. Germany: LAP LAMBERT Academic Publishing; 2015. p. 76

[4] Engebretson DC, Cox A, Gordon RC. Relative motions between oceanic and continental plates in the Pacific Basin. Geological Society of America Special Paper. 1985;206:1-59

[5] Otofuji Y, Hayashida A, Torii M. When was the Japan Sea opened? Paleomagnetic evidence from Southwest Japan. In: Nasu N, Uyeda S, Kushiro I, Kobayashi K, Kagami H, editors. Formation of Active Ocean Margins. Tokyo: Terra Publishing Co.; 1985. pp. 551-566

[6] Hall R. Cenozoic geological and plate tectonic evolution of SE Asia and the SW Pacific: Computer-based reconstructions, model and animations. Journal of Asian Earth Sciences. 2002;20:353-431

[7] Itoh Y, Nagasaki Y. Crustal shortening of Southwest Japan in the Late Miocene. The Island Arc. 1996;5:337-353

[8] Itoh Y, Maekawa H. A Story of Mountain Building: Neotectonic Evolution of a Convergent Margin on the

Northwestern Pacific. New York: Nova Science Publishers, Inc.; 2021. p. 66

[9] Nakamura K, Renard V, Angelier J, Azema J, Bourgois J, Deplus C, et al. Oblique and near collision subduction, Sagami and Suruga Troughs— Preliminary results of the French-Japanese 1984 Kaiko cruise, Leg 2. Earth and Planetary Science Letters. 1987;83:229-242

[10] Makimoto H, Miyata T, Mizuno K, Sangawa A. Geology of the Kokawa District, with Geological Sheet Map at 1:50,000. Tsukuba: Geological Survey of Japan, AIST; 2004. p. 89

[11] Noda A, Toshimitsu S. Backward stacking of submarine channel-fan successions controlled by strike-slip faulting: The Izumi Group (Cretaceous), southwest Japan. Lithosphere. 2009;1: 41-59

[12] Noda A. Strike-slip basin—Its configuration and sedimentary facies. In: Itoh Y, editor. Mechanism of Sedimentary Basin Formation—Multidisciplinary Approach on Active Plate Margins. Rijeka: InTech; 2013. DOI: 10.5772/56593

[13] Takahashi J, Yamasaki T, Yokota Y, Kawanishi J, Inoue S. Geology of the Iyo City—Futami Town area, Ehime Prefecture. Memoirs of Faculty of Education (Natural Science), Ehime University. 1990;10:19-29

[14] Geological Survey of Japan, AIST, editor. Seamless Digital Geological Map of Japan 1:200,000 (May 29, 2015 Version), Research Information Database DB084. Tsukuba: Geological Survey of Japan, AIST (National Institute of Advanced Industrial Science and Technology); 2015

[15] Nakata T, Imaizumi T, editors. Digital Active Fault Map of Japan with 2 DVDs. Tokyo: University of Tokyo Press; 2002. p. 60

[16] Miyazaki K, Wakita K, Miyashita Y, Mizuno K, Takahashi M, Noda A, et al. Geological Map of Japan 1:200,000, Matsuyama. 2nd ed. Tsukuba: Geological Survey of Japan, AIST; 2016

[17] Okada A, Tsutsumi H, Nakata T, Goto H, Niwa S. Active Fault Map in Urban Area 1:25,000, Gunchu. Tsukuba: Geospatial Information Authority of Japan; 1998

[18] Takahashi J, Kashima N. On the Gunchu Formation at the Mori Coast Iyo City, Ehime Prefecture. Memoirs of Faculty of Education (Natural Science), Ehime University. 1985;5:19-29

[19] Mizuno K. Preliminary report on the Plio-Pleistocene sediments distributed along the Median Tectonic Line in and around Shikoku, Japan. Bulletin of the Geological Survey of Japan. 1987;38:171-190

[20] Kitabayashi E, Danhara T, Iwano H. Fission-track age of zircon from volcanic ash layer of the Gunchu Formation in Iyo City, Ehime Prefecture in Shikoku, Japan. Journal of the Geological Society of Oita. 2012;18:61-64

[21] Itoh Y, Iwata T. Structural features along the Median Tectonic Line in southwest Japan: An example of multiphase deformation on an arc-bisecting fault. In: Itoh Y, editor. Evolutionary Models of Convergent Margins—Origin of their Diversity. Rijeka: InTech; 2017. DOI: 10.5772/67669

[22] Kusumoto S, Fukuda Y, Takemura K, Takemoto S. Forming mechanism of the sedimentary basin at the termination of the right-lateral left-stepping faults and tectonics around Osaka Bay. Journal of Geography. 2001;110:32-43

[23] Itoh Y, Kusumoto S, Takemura K. Characteristic basin formation at terminations of a large transcurrent fault—Basin configuration based on gravity and geomagnetic data. In: Itoh Y, editor. Mechanism of Sedimentary Basin Formation—Multidisciplinary Approach on Active Plate Margins. Rijeka: InTech; 2013. DOI: 10.5772/56702

Section 2

Geophysical Methods in Geological Applications

Chapter 4

Tectonic Collision, Orogeny and Geothermal Resources in Taiwan

Chao-Shing Lee, Lawrence Hutchings, Shou-Cheng Wang, Steve Jarpe, Sin-Yu Syu and Kai Chen

Abstract

The recent tectonic evolution of Taiwan created ideal conditions for geothermal resources: heat, water and permeability. We examine heat flow measurements, seismic tomography, seismicity, hot spring distribution, tectonic history, geology, and volcanism described in previous studies to understand the relation between tectonics and geothermal potential in Taiwan. Taiwan is the youngest tectonically created island on earth. The island formed as a result of the transition from subduction of the Eurasian Plate under the Philippine Sea plate to active collision. Collision results in orogenic mountain building. The geology of the island is primarily an accretionary prism from the historic subduction. This active orogeny creates unusually high geothermal gradients by exhumation of the warmer material from depth and by strain heating. As a result, temperatures reach up to ~200 degree C. Volcanoes in the northern tip of Taiwan provide an additional source of heat. Favorable fluid flow from meteoric waters and permeability from seismicity and faulting results in exploitable geothermal systems near the surface. These systems can potentially provide geothermal power generation throughout the whole island, although there are currently only two geothermal power plants in Taiwan.

Keywords: geothermal, Taiwan tectonics, Taiwan heat

1. Introduction

Most of the energy consumed in Taiwan is produced by imported fossil fuels. The development of renewable energy is a priority, both to move the country toward energy independence and to combat global climate change. Power produced from geothermal energy is an important component of a renewable energy portfolio because it is continuously available, as opposed to wind and solar, which are intermittent in availability.

The presence of numerous hot springs and high measured heat flow throughout Taiwan are indications that significant geothermal energy resources are present, even though only two geothermal power plants are currently operating in the country. **Figure 1** shows the distribution of hot springs. The hot springs and high heat flow at the northern tip of Taiwan are associated with known volcanoes, but, interestingly, thousands of hot springs are also found in many other areas from the north to the south where no volcanism is found. The potential for geothermal

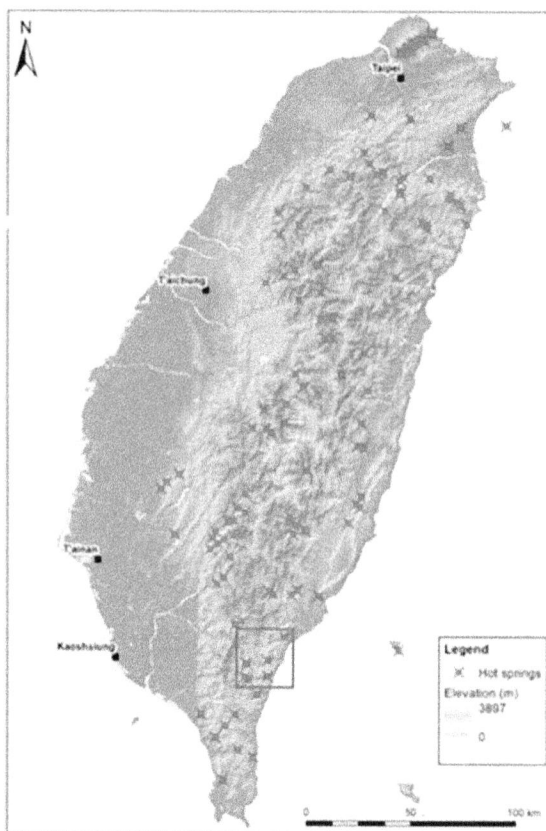

Figure 1.
Hot springs of Taiwan. Light beige indicates the central range, with foothills to the east and the Longitudinal Valley (a long green strip) east of the foothills. From [1, 2] the blue box identifies the area of the highest heat flow in Taiwan.

energy from these sources is perhaps more relevant for Taiwan than the limited volcanic area, as their spatial distribution is much wider and thus can perhaps serve a much larger area.

In this chapter we examine previous studies to show the relation between evolutionary tectonics and the potential for geothermal resources in Taiwan. We show how this evolution created ideal conditions for geothermal resources; heat, water and permeability. We examine heat flow measurements, seismic tomography, seismicity, hot spring distribution, tectonic history, geology, and volcanism described in previous studies to understand the relation between tectonics and geothermal potential in Taiwan.

2. Evolutionary tectonics

Prior to 6–9 Ma, sea floor spreading occurred in the South China Sea and the lithosphere within the latitudes of Taiwan was subducted eastwards beneath the Philippine Sea Plate along the Manila Trench [3]. Convergence was at a 5.6 cm/year with respect to Eurasia [3, 4]. The convergence created an accretionary prism of varying ages, a result of off-scraping sediments and rocks from the under-thrust plate and depositing sediments and volcanic ashes from volcanic arcs on the over-riding plate [5, 6]. The sediments are a result of marine sedimentation occurring on

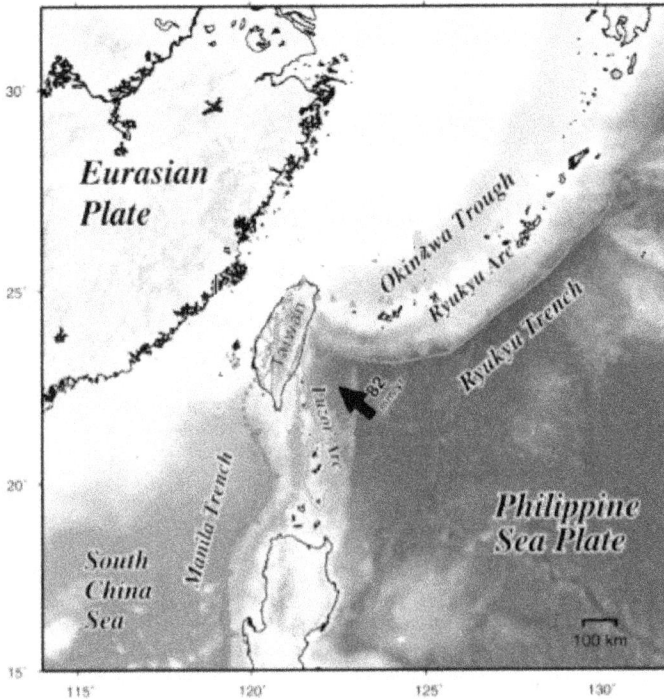

Figure 2.
Major tectonic features near Taiwan (modified from [10]). Large red triangles indicate direction of subduction. Small red triangles along the north of the island are the volcanoes of the active volcanic front of the Ryukyu arc [11]. Volcanoes to the southeast are part of the Luzon arc.

the South China Sea oceanic crust during the Miocene [5, 7]. Taiwan is an accretionary prism from the previous subduction [8]. About 6–9 Ma sea floor spreading ceased in the South China Sea leaving the Eurasian plate subducting eastward under the Philippine Sea plate. At the same period the Luzon arc acquired a significant enough topographic expression to resist subduction and start to collide with the Eurasian plate. The sediment strata began to show evidence of plate collision in early Pliocene and rising of Taiwan, about 5 Ma [8, 9]. Active subduction continues today north and south of Taiwan (**Figure 2**).

The Luzon volcanic arc (LVA) is an intra-oceanic volcanic arc which belongs to the Philippine Sea plate (PSP). Today the LVA is in contact with Coastal Range of Taiwan, along the Longitudinal Valley in south western Taiwan [3, 4, 12]; **Figures 2** and **3**. Fission track dating in the Central Range indicates a gradual rising until about 2 Ma when it began to accelerate [2, 4] and caused the Central Range to experience rapid uplift of the roughly 40 km wide range for most of the length of Taiwan; the highest mountain uplift in the world, as much as ~26 mm/year. Other regions experiencing present-day uplift in Taiwan at rates of ~0.2–25.8 mm/yr. include the Hsuehshan Range, the Central Range, and the southern part of the Western Foothills [14, 15] **Figure 4**. An uplift rate of 22.9 mm/yr. occurs in the central northern portion of the Central Range, and the southern part of the Coastal range displays uplift rates of ~1.3–25.8 mm/yr., the largest in Taiwan. The uplift rates decrease rapidly toward the north and diminish gradually toward south [17].

A number of studies agree oblique collision results in the evolution through time of Taiwan mountain building visible as a continuum from the present-day Manila

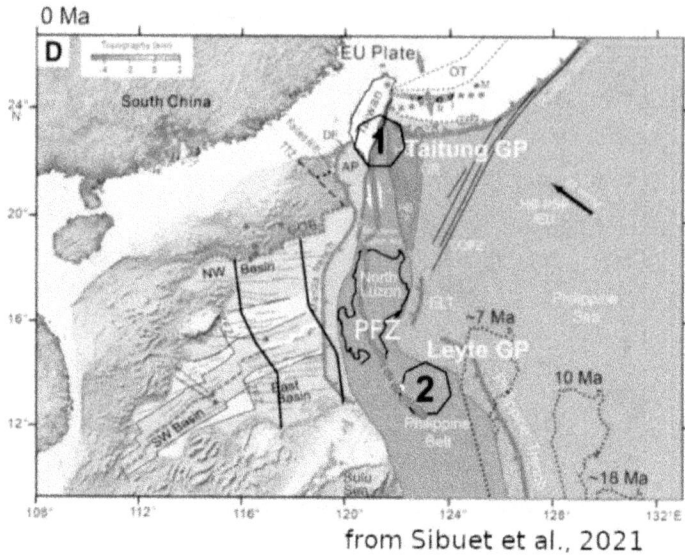

from Sibuet et al., 2021

Figure 3.
From [13]. Thick red lines are faults. Triangles point to subduction, plain line is the Philippine fault hypothesized to extend into Taiwan. Description of other feature are available in [13] the Taitung (1) and Leyte (2) geothermal prospects are located along the left-lateral strike-slip Philippine fault [13]. Black dots shown the location of the Luzon volcanic arc.

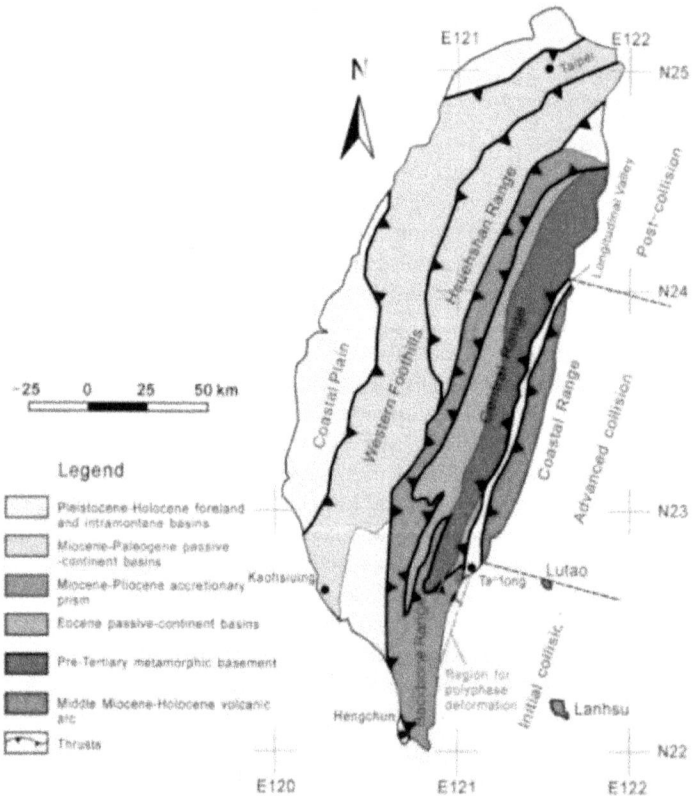

Figure 4.
Provinces (from [16]).

subduction system to the south (before collision), through middle Taiwan (collision) and northeast of Taiwan, across the southern Okinawa Through and Ryukyu subduction system (post-collision) [3, 4, 10]. Also, the age of metamorphic geology increases from the west to the east across Taiwan [18].

There are many thrust faults and folded anti- and syn-clines roughly perpendicular to the convergence. Active seismicity and highly fractured zones create permeability. The 1999 M = 7.8 Chi-Chi earthquake occurred on a thrust fault in the Central Range and demonstrates one way stress buildup from collision is released [19, 20]. The Longitudinal Valley is oblique to the convergence and hosts a left-lateral strike-slip fault, which may be an extension of the Philippine Fault that runs through the Luzon and several smaller islands of the Philippines (**Figure 3**). The saturated fractured geology creates the pathways for hot water and steam to come up to the surface making the Taitung and Leyte geothermal prospects (**Figure 3**). Sibuet et al. [13] connects this fault with the Longitudinal left-lateral fault. In Taiwan, most geothermal prospects are located along this key fault.

3. Geology

Seven provinces of the Taiwan orogen can be recognized (**Figure 4**): (1) the Coastal Range, along the northern extension of Luzon arc, consisting of fore arc sedimentary units and andesitic volcanic rocks; (2) the Longitudinal Valley, filled with young sediments, being the plate boundary between the Eurasian and Philippine Sea Plates; (3) the eastern Central Range, consisting of pre-Tertiary (Tailuko belt) and Miocene (Yuli belt) metamorphic complex rocks; (4) the western Central or the Backbone Range (capped by Miocene slates), (5) the Hsuehshan Range (Eocene-Miocene slates); situated in the northern half of the island and tapers off toward the south, (6) the Western Foothills, the fold-and-thrust belt, composed of clastic Oligocene-Pleistocene sedimentary rocks, and; (7) the Coastal Plain, containing younger sediment deposits [16, 21]. Longitudinal Valley represents the tectonic suture zone separating metasedimentary sequences of the Central Ranges from the accreted sedimentary and volcanic arc rocks of the eastern Coastal Range.

Continental crust of the Chinese continental margin colliding with the Luzon Volcanic Arc has deformed Miocene to Quaternary sedimentary marine stratigraphy into easterly-dipping fold and thrust belts of the Western Coastal Plains, Western Foothills and the Hsuehshan Range in western Taiwan Island. To the east a metamorphosed continental margin sequence has been exhumed along westerly dipping faults as the Central Ranges. Along the eastern side of Taiwan Island, the Coastal Range represents the northern extent of the Luzon Volcanic Arc, which has been accreted onto the eastern margin of the exhumed metamorphic rocks. The geology of the accretionary prism of the eastern Central Range is primarily made up of a huge sequence of deep-sea turbidities (Miocene Lushan Formation). It consists of mudstones, siltstones, and sandstones [16, 22].

4. Heat

In Taiwan, in the mature collision zone, the heat flows are high. Heat flow in the mountains is mostly between 80 and 250 mW/M^2 [23]. Overall, the high heat flow in the mountainous regions is interpreted as a result of rapid uplift and exhumation of the warmer material from depth [24–26]. However, several researchers in orogenic zones have shown that uplift alone is not sufficient to account for the high heat flow [27–30]. Theoretical and geological considerations suggest that viscous heating is a cumulative process that may explain the heat deficit in collision oro-gens; where severely deformed rocks over a short time span cause viscous heating and can account for this deficit and explain further up-warping of the isotherm [31, 32]. Whereas radiogenic heat production can be inferred from measured concentrations of radioactive elements and heat flow in stable regions of the litho-sphere, the contribution to heating by deformation can potentially be measured only in actively deforming lithosphere where it may not be easily separated from other sources of heat. The strain heating and upwarping together, create favorable geothermal gradients in the Central mountains of Taiwan. The shallowest isotherm in the Central Range may also account for the aseismic zone, likely due to too hot and pliable crust unable to sustain enough stress to generate earthquakes [8]. Other sources near surface heat may include groundwater circulation, topographic effects, and higher radiogenic heat production rates in the continental crust. Complex deformations in lower crust and upper mantle following the collision might have also affected the thermal structures in this region [33].

Heat flow gives valuable insight into evaluating the tectonics of a region, and the geothermal gradient is the primary initial indicator of a viable geothermal resource. Heat flow is derived from the geothermal gradient and thermal conductiv-ity, which are typically obtained from temperature measurements in pre-existing wells or laboratory samples. Heat flow is generally described in units of milli Watts per meter squared (mW/M^2), and the geothermal gradient is described as degrees Celsius per kilometer (°C/km).

Figure 5 shows heat flow measurements from [25]. They studied heat flow in Taiwan using previous values obtained from [27, 33] and added many values

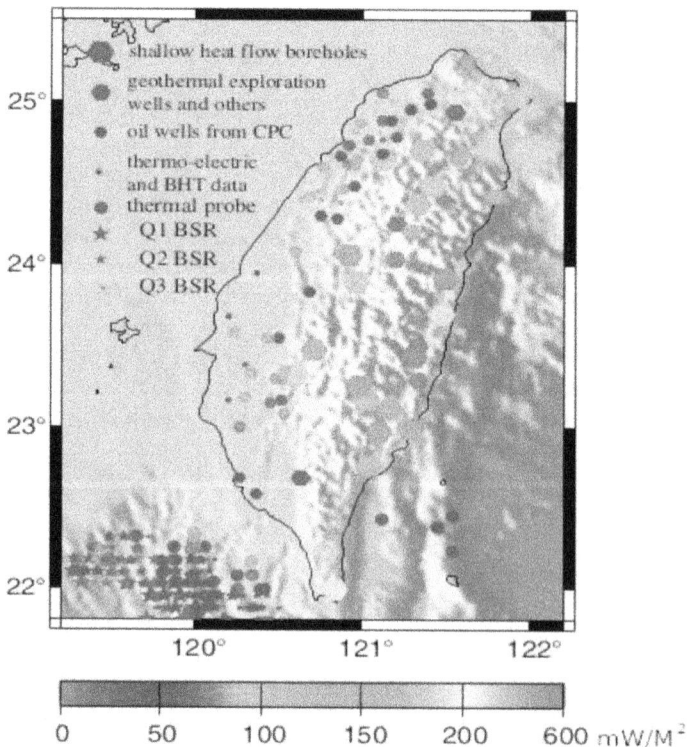

Figure 5.
Heat flow measurements from [33]), re-plotted from [25].

obtained off-shore of Taiwan. Chi and Reed [25] points out that there is still debate whether the heat flow data from some of the "geothermal wells" are representative of the regional heat. However, they also point out that several studies are able to fit the high heat flow pattern by thermal modeling using different crustal kinematic models [33–35]. Within Taiwan, and along the central range, heat flow reaches over 300 mW/M^2, whereas the worldwide average is about 50 mW/M^2.

Chi and Reed [25] identified a dramatic difference in heat flow between the subduction zone to the south, where values are near world averages, and the collision zone in Taiwan. At a latitude of ~20.5°N (not shown in **Figure 5**), in the subduction zone, heat flow decreases even further from about 75 to 40 mW/M^2 from the trench to the upper slope domain of the accretionary prism. To the east in the fore arc basin, heat flow values are ~25 mW/M^2 (**Figure 5**). The heat flow pattern along this transect is consistent with the three in situ heat flow measurements farther to the south at ~19°N [33, 35]. Heat flows in the satellite basins in the arc region are ~50 mW/M^2. Farther to the north in the initial collision zone (**Figure 5**), the continent-ocean boundary (COB) enters into the trench near 21.2°N. Hwang and Wang [36] have collected 12 thermal probe measurements along a transect from continental shelf (117°E, 22.8°N) to continental slope (118.1°E, 19.3°E) that is 220 km west of and parallel to the trench. Chi and Reed [25] treat this data set as the initial condition before the Chinese passive margin enters into this convergent boundary. Hwang and Wang [36] found that heat flows are ~80 mW/M^2 in the continental slope and decrease to 70 mW/M^2 in the abyssal plain. Chi and Reed [25] also found slightly increased heat flows once the incoming sediments were scraped off and incorporated into the toe of the accretionary prism, where intensive dewatering occurs. High geothermal

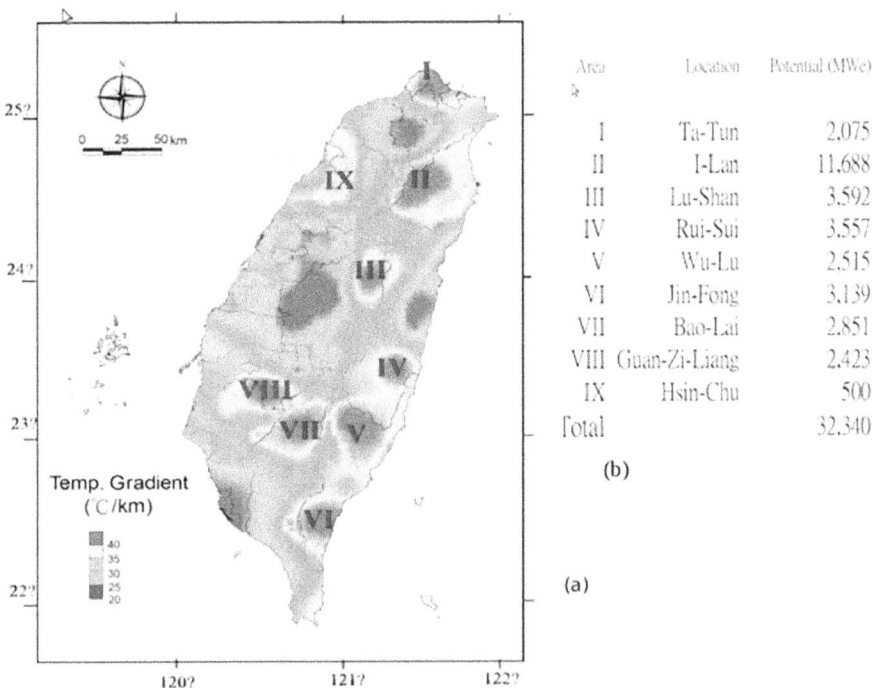

Area	Location	Potential (MWe)
I	Ta-Tun	2.075
II	I-Lan	11.688
III	Lu-Shan	3.592
IV	Rui-Sui	3.557
V	Wu-Lu	2.515
VI	Jin-Fong	3.139
VII	Bao-Lai	2.851
VIII	Guan-Zi-Liang	2.423
IX	Hsin-Chu	500
Total		32.340

(b)

(a)

Figure 6.
a. Temperature gradient map of Taiwan showing nine EGS regions with anomalous high temperature gradients: (I) Tatun; (II) Chingshui-Tuchang; (III) Lushan; (IV) Juisui-Antung; (V) Wulu-Hungyeh; (VI) Chihpen-Chinlun; (VII) Paolai; (VIII) Kuantzuling; (IX) Hsinchu-Miaoli. b; potential Mwe of the nine zones. From [37].

gradients (40–80°C/km) and heat flows (50–105 mW/M^2) were found in a thick basin near the toe in this region east of the COB (**Figure 2**). This might be a result of intensive dewatering in this basin, which covers a circular region with a diameter of 60 km centered at 119.8°E, 21.6°N. Seismic reflection data show conjugate fault plane reflections within the basin, even though the displacements across the faults are small, suggesting possible fluids within the fault zones. Away from the toe in the initial collision zone, the heat flows decrease toward the arc as the sediments stack thicken, especially from lower slope domain to the upper slope domain. Heat flows ranging from 30 to 60 mW/M^2 and increasing toward the arc are identified in the back-thrust domain.

Taiwan has a very good geothermal gradient. The average land heat flow in the world is about 30 mW/M^2. If there is a local heat flow value greater than the average, then there is a good potential for geothermal development [37]. Song and Liu [37] identified nine major geothermal resource regions based on anomalous geothermal gradients >35°C/km (**Figure 6**): (I) Tatun volcanic area in the north; (II) Yilan Plain along the Lishan Fault, it extends southwest to Mount Lu, covering Jiaoxi, Qingshui, Tuchang, Lushan and other geothermal areas; (III) For regions with abnormal geotemperature gradients higher than 35° C/km Lushan; (IV) Ruisui-Antong; (V) Wulu-Hongye; (VI) Zhiben-Jinlun; (VII) Baolai; (Viii) Guanzailing; (IX) Hsinchu-Miao Li. These nine high anomalous areas of geothermal gradient are the potential areas of geothermal development. Taiwan is currently conducting small-scale exploration to potentially exploit these areas.

5. Geothermal reservoirs

Organic geothermal reservoirs generally are permeable zones with fluid flow and heat. Taiwan has about 2500 mm of rain each year [38], sufficient for replenishing reservoirs. Engineered geothermal reservoirs only require heat, and fluid and permeability are created artificially. Heat is the important source for geothermal development in both cases. In Taiwan, orogenic reservoirs are typically located at a shallow depth of 2–3 kilometers in the Central range and volcanic zone where the geothermal gradient is greater than 30° C/km. The development depth of enhanced geothermal away from the Central range is usually greater than 3000 m where the geothermal gradient is approximately 30 mW/M^2. Exploration is necessary to determine whether there is an organic reservoir or a candidate for an engineered system. Within areas outlined in **Figure 6**, there are 108 geothermal potential locations for geothermal development that are located at a shallow depth of 2–3 kilometers due to high geothermal gradients [39]. Except for the Tatun and Keelung Volcano Groups in the north, the main high-gradient areas are located in the metamorphic rock belts.

The Tatun volcanic group is located in the northernmost part of Taiwan, mainly composed of more than 20 Pleistocene andesite volcanoes such as Tatun Mountain. At Tatun, there is a classical lava cone, and there are many hot springs, fumaroles, sulfur pores [40, 41], and other indications of intense geothermal activity. The main shallow geothermal reservoirs are located between the Jinshan and Kanjiao faults. The temperature is about 200-290° C [40]. Deep geothermal wells have encountered a high temperature of 293° C, which is the highest temperature currently reached in Taiwan [25].

Organic geothermal reservoirs can be roughly divided into two types: hot water type and steam type. Engineered geothermal system (EGS) is another type of reservoir. Organic geothermal energy development is limited by hydrothermal production capacity. In geothermal fields or reservoirs, abundant and high-temperature geothermal water and well-developed fissure structures are required. The hot water type geothermal system is based on the presence of hot water in the reservoir. The water phase controls the reservoir pressure, and its temperature is the highest. Temperatures can range from less than 100° C to 370° C, but geothermal system above 200° C are optimum for power generation. The vapor type of reservoir is formed by the high temperature heat source supplying heat and the low permeability of the rock formation. Reservoirs can evolve from water type to steam type if the amount of hot water extracted is more than the amount of that replenished from groundwater. In the vapor reservoir, two phases of hot water and vapor coexist, and the vapor phase controls the storage layer pressure. The hot water phase flows in small pores due to high surface tension, while the vapor phase escapes through larger conduits in the upper geothermal system. Currently, steam type reservoirs only account for 10% of global geothermal production.

Engineered geothermal system typically create permeability between two wells by hydro-fracking or controlled fracturing from injection of cold water. **Figure 7** shows a scheme of an engineered system. This approach has been largely unsuccessful to date. This is because one, the fluid is not heated sufficiently in the short distance traveled, and two there has been difficulty creating a large enough fracture system, which would help in solving the first problem. Because the general gradient is about 30° C/km, so the development depth of enhanced geothermal is usually greater than 3000 m.

According to the geothermal research report [42], if 2% of the thermal energy stored in the 3–10 km rock layer of the earth's crust can be obtained the heat energy produced by it is as high as 2.8 × 10 5 EJ, which is about 2800 times the

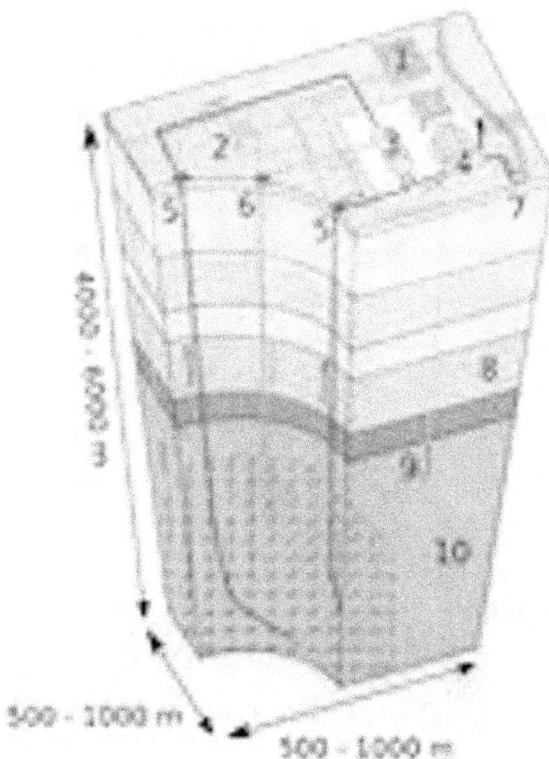

Figure 7.
Engineered geothermal reservoir.

total energy consumption of the United States in 2005. There is a huge geothermal potential for engineered geothermal systems deep underground in Taiwan. If it can be successfully developed, it will be an important independent energy source in Taiwan.

Two geothermal power plants, a total of 6 MWe, have started to produce electricity in Taiwan at the end of 2021. Two other small geothermal plants, 1 and 2 MWe, are being developed in indigenous areas, and are planned to operate by the end of 2022. The government has established a short-term geothermal goal to reach 200 MWe by 2025 with expansive exploration, test drilling, and power plant management. Further, the government is investing in developing engineered geothermal systems to reach long-term goals. It is hoped Taiwan may reach the GWe-level geothermal power production by 2035, thereby reducing the carbon dioxide emission as part of a global-village member.

6. Methods for exploration

Geophysical methods for exploring for orogenic geothermal reservoirs include the gravity, aeromagnetic, electro-magnetic (magnetotelluric), and tomographic imaging. The gravity method models the spatial distribution of rocks with different densities at depth to match the measurements of the acceleration of gravity at

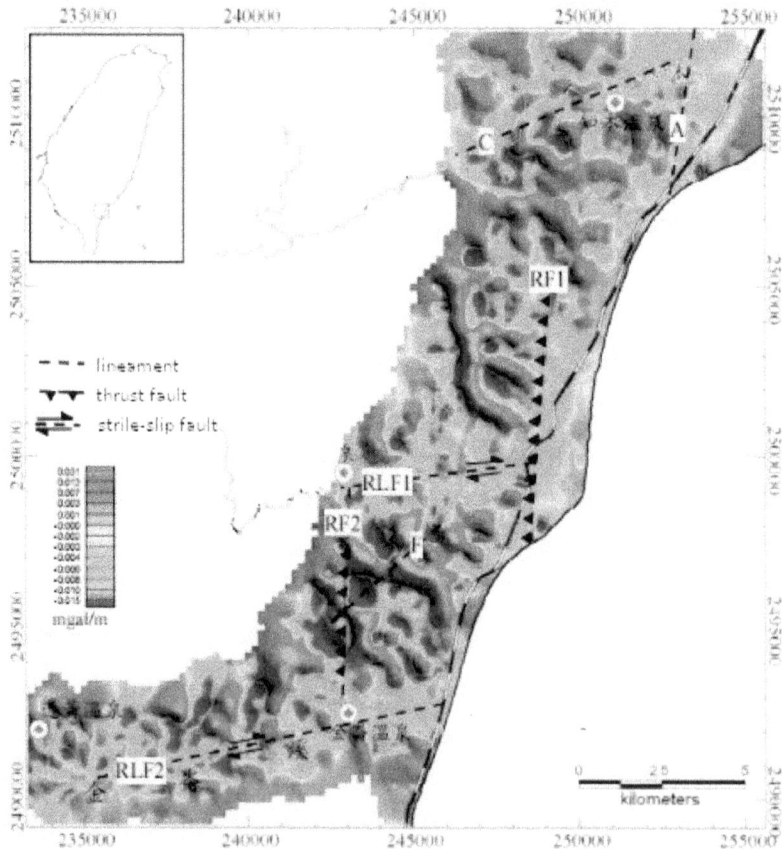

Figure 8.
*Gravity showing low density rocks with geothermal potential at RF1, RFL1, RF2 and RFL2 along faults. Square in sub-figure shows the area, which is the same as the square showing high heat flow in **Figure 1**.*

different positions on the surface. **Figure 8** shows contours of gravity in mgals in southern Taiwan [43]. Geothermal reservoirs typically occur at low gravity values as seen below RF1, RFL1, RF2 and RFL2 where the gravity contours show significantly lower values than surrounding rocks.

An aeromagnetic survey records the magnetic field from an air plane flying over the area of interest. Magnetic field measurements are typically at 500 m intervals and the survey area is crisscrossed in parallel and perpendicular directions. Typically, a helicopter is flown over at a height of about 500 m above topography. In addition, magnetometers are used to continuously measure the geomagnetic field at a chosen base station on the ground. Normally, correction for the International Geomagnetic Reference Field (IGRF) involves removing it in order to show only magnetic anomalies related to geology. Resolution can be as high as 100 m from the surface to 10 km depth [44], **Figure 9** shows typical results over a geothermal reservoir. In this case, the linear features of the magnetic susceptibilities (H1, H2, and H3) were interpreted as being separated by large linear vertical faults. Although there is no direct evidence of a geothermal reservoir in this particular aeromagnetic survey, the delineation of structures such as faults can aid in the search process.

Electromagnetic methods can be either passive, utilizing natural ground signals (e.g. magnetotellurics) or active, where an artificial transmitter is used either in the near field (as in ground conductivity meters) or in the far field (using remote high powered military and civil radio transmitters as in the case of VLF and RMT methods). Magnetotelluric (MT) surveys estimate the Earth's electromagnetic impedance by measuring naturally occurring electromagnetic waves in a very broad frequency range. They typically record the full component MT data (i.e., Ex, Ey, Hx, Hy and Hz) induced by natural primary sources and measured relatively uniformly at 1000 m

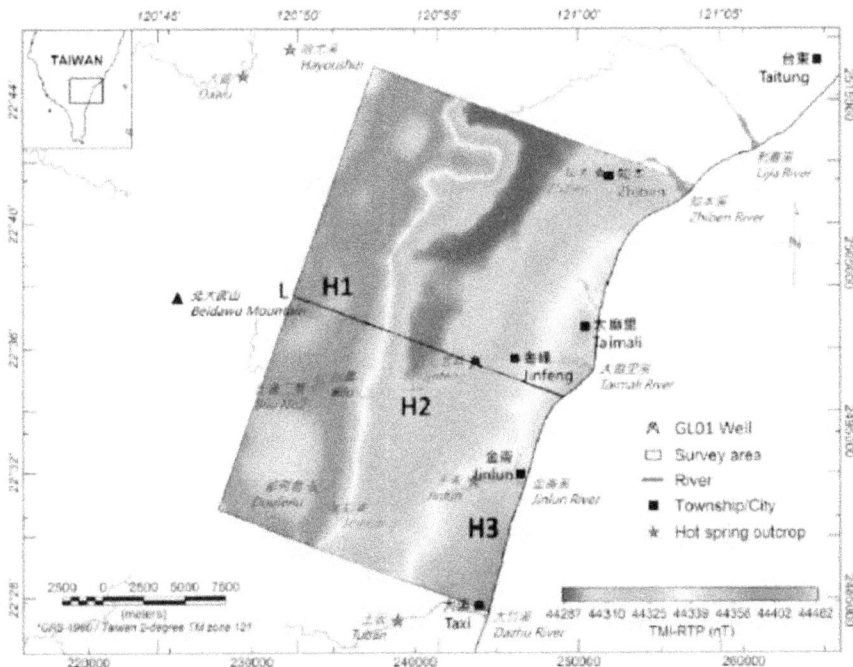

Reduced to the TMI map from Lin and Tong (2020). H1, H2 and H3 are NS trending high magnetic belts interpreted to be separated by left-lateral faults.

Figure 9.
Magnetic susceptibilities. From [44].

intervals across the area of interest. Frequency range is generally from 10 kHz down to 0.01 Hz and can be even lower when sounding duration is as long as five days. Interpretation is based upon inversion of the MT data to derive resistivity values. The final 3D model used for interpretation is the one with the lowest root mean square misfit. **Figure 10** shows typical MT results at three depths in a reservoir in southern Taiwan. The low resistivity values (green) can indicate the location of a reservoir. The underground resistivity is very low resistance layer of 10 Ω-m. When the rock layer is subjected to hot water, the resistivity of the formation will be significantly reduced. Further abnormally low (below 100 Ω-m) area may indicate a heat source.

A recently developed seismic method utilizes passive earthquake sources and dense recording networks to image reservoirs, and has shown promise in seismically

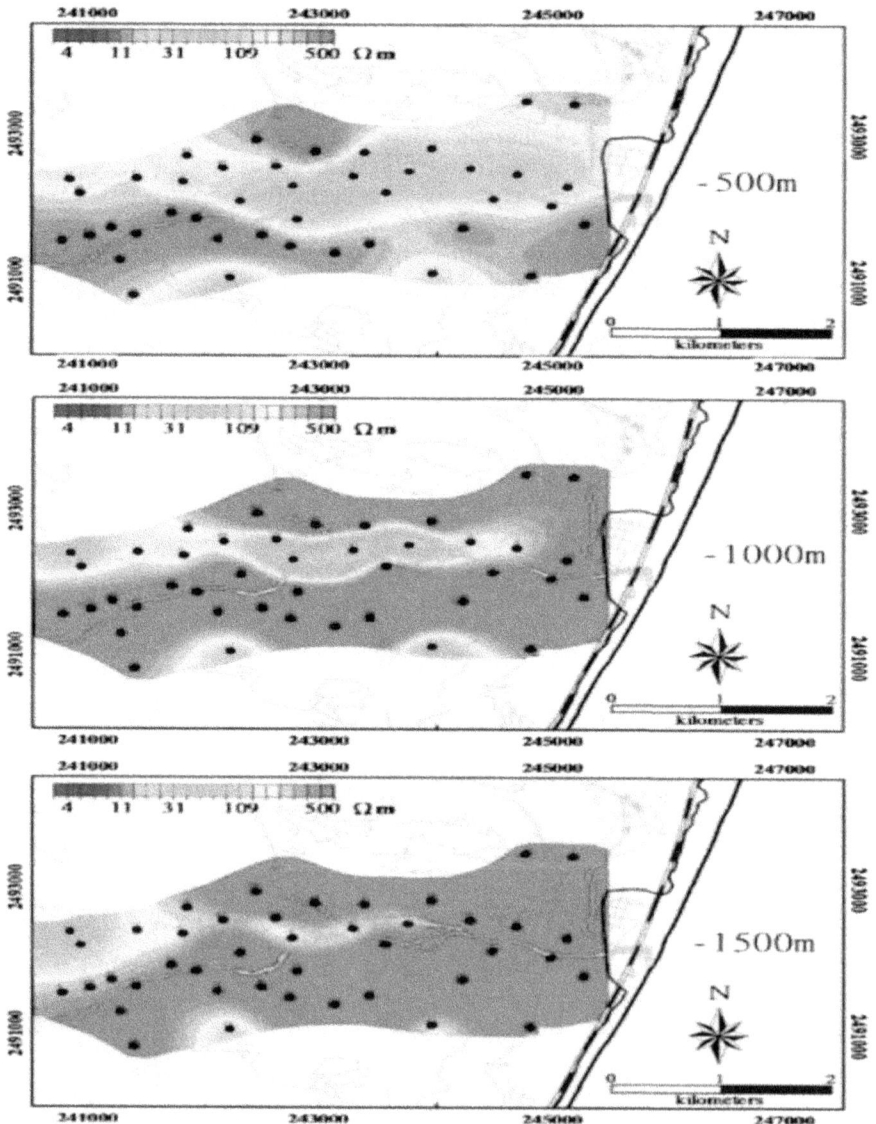

Figure 10.
Resistivity values at three depths. From [45].

active areas [46, 47]. Inexpensive recorders and automated data processing makes this possible in a short amount of time at a minimal cost [48]. The propagation energy from the earthquakes passes through the geology to recording systems at the surface. Tomographic inversion is used to back project these recordings to provide the images at depth [49]. The primary information comes from propagation of first arriving *P*- and *S*-waves and their pulse widths. These provide *P*- and *S*-wave velocity and Qp and Qs attenuation structure throughout the volume [46]. Attenuation measures the energy lost as waves propagate through the geology.

Further, the values of *Vp* and *Vs* throughout the volume can be used to calculate Poisson's ratio, and Bulk, Young's, Lambda and shear moduli throughout the

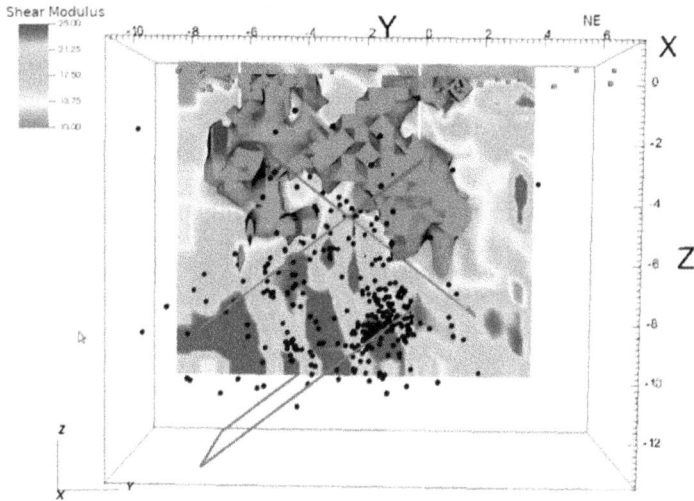

Figure 11.
Shear modulus. Low values likely indicate fractures. From [43].

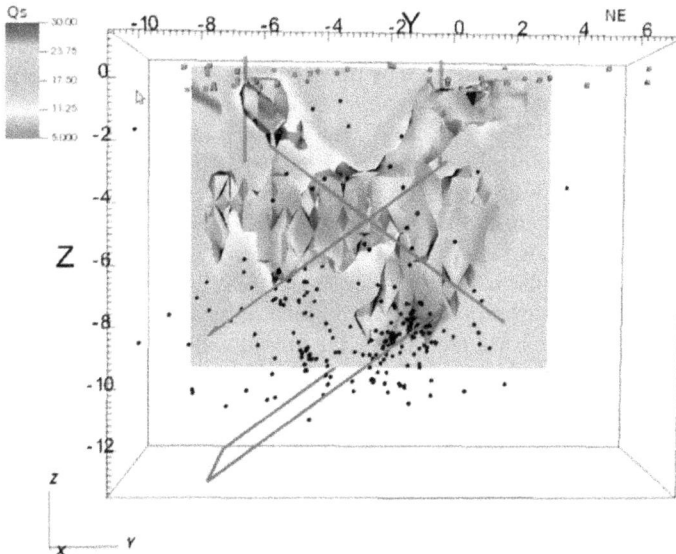

Figure 12.
Qs. Low values (orange) likely indicate saturation. From [45].

volume. These proprieties are utilized in the context of rock physics relationships to identify the effects of fluids, fractures, porosity, and permeability on seismic velocity [46]. Following [46], several typical interpretations of porosity, permeability and saturation can be made from observable microearthquake data. Comparisons are made relative to normal geology at similar depths and temperatures, meaning geology that has a monotonic increase in velocity and Q as a function of depth, and saturation, porosity and temperature that is considered average for the geologic condition of the study area [46].

Figure 11 shows a cross-section of a likely reservoir in southern Taiwan [43]. Black dots are micro-earthquakes. Low values of shear modulus (red) indicate a soft geology [43], likely due to fractures and may indicate the location of the reservoir. Almost no earthquakes occur in the soft geology. This shows how faulting can create permeability. The faults create pathways for fluids, and highly fracture the geology. Here, the conjugate faulting matches the flanks of anticlines caused by the orogeny deformation [43].

Figure 12 shows Q_S for the same volume [45]. The low values of Q_S can indicate the existence of water. Shear wave propagation is not affected by liquid, so this figure is independent of the shear modulus. Together, the fractures and saturation are a good indication of a reservoir. Blue lines indication the location of possible faults, which may provide pathways for water.

7. Induced seismicity

During EGS geothermal reservoir creation or organic stimulation, rocks may slip along pre-existing fractures and produce microseismic events. Researchers have found these microseismic events, also known as induced seismicity, to be a very useful diagnostic tool for accurately pinpointing where fractures are re-opened or created, and characterizing the extent of a reservoir. In almost all cases, these events occur in deep reservoirs and are of such low magnitude that they are not felt at the surface. Although induced seismicity data allows better subsurface characterization, GTO also understands public concern. With this in mind, US DOE led an effort to create a protocol for addressing induced seismicity associated with geothermal development, which all US DOE-funded EGS projects are required to follow [50]. This work was informed by panels of international experts and culminated in an International Energy Agency- accepted protocol in 2008. The protocol was updated in early 2011 to reflect the latest research and lessons learned from the geothermal community. In addition, in June 2012, the US National Academy of Sciences (NAS) issued Induced Seismicity Protocol in Energy Technologies [51]. The report found that geothermal development, in general, has a low potential for hazard from induced seismicity. The NAS report cited the DOE Induced Seismicity Protocol as a best practice model for other subsurface energy technologies.

Figure 13 is a demonstration of induced seismicity due to cold water released by gravity flow over a month period of time in four wells at the Geysers, California geothermal reservoir [52]. The white dots are the earthquakes, including the natural and human-induced earthquakes. V_S tomography results are shown as backdrop. In **Figure 13a**, the well at far left, water release was increased just prior to the month. For the well second from left, water was released only for two days at the end of the month. Seven earthquakes occurred after water release was started and no earthquakes occurred prior. For the well third from left water was released at a low rate during the month. The seismicity was relatively little near the bottom of the well. The well at the far right is not in hot geology, and no earthquakes are observed.

Injection at four wells

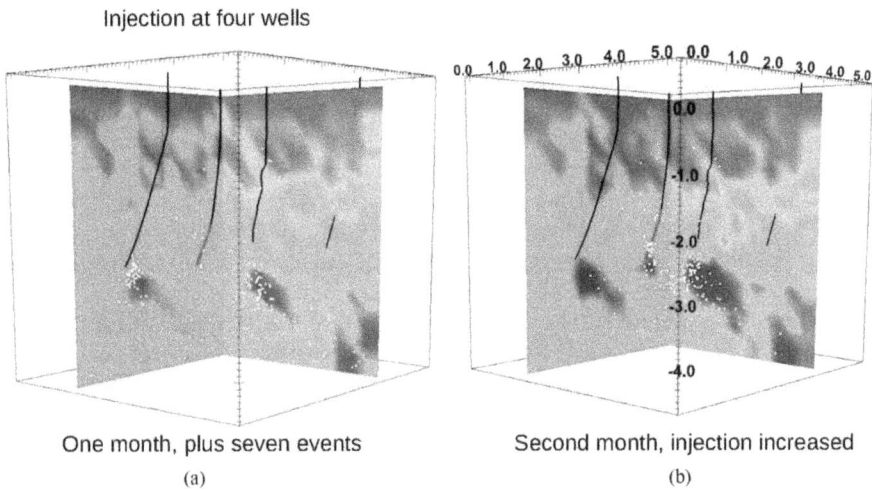

One month, plus seven events | Second month, injection increased
(a) | (b)

Figure 13.
Demonstration of induced seismicity due to the water release in hot geology.

Figure 13b shows the effect of water injection at four wells for the month immediately following. The well on the far left has ceased generating earthquakes. The well second from left has created many induced earthquakes due to the ongoing release of water. The third well from the left has increased water flow and shows an increase in seismicity at the bottom of the well. The well at the far right continues to show no earthquakes.

The case study presented above illustrates that by continuously monitoring seismic activity and modifying water injection accordingly, the occurrence of induced seismicity can be mitigated.

8. Discussion and conclusions

Power produced from geothermal energy is an important component of a renewable energy demand because it is continuously available, as opposed to wind and solar, which are intermittent in availability.

The recent tectonic evolution of Taiwan created ideal conditions for geothermal resources; heat, water and permeability. The island formed as a result of the transition from subduction of the Eurasian Plate under the Philippine Sea plate to active collision. The Central Range has been the most dramatic manifestation of this collision. It has experienced the highest mountain uplift in the world, as much as ~26 mm/year. This active orogeny creates unusually high geothermal gradients by exhumation of the warmer material from depth and by strain heating. As a result, temperatures reach up to ~200 degree C. Volcanoes in the northern tip of Taiwan provide an additional source of heat. Favorable fluid flow from meteoric waters and permeability from seismicity and faulting results in exploitable geothermal systems near the surface. These systems can potentially provide geothermal power generation throughout the whole Island.

The average land heat flow in the world is about 30 mW/M^2. If there is a local heat flow value greater than the average, then there is a good potential for geothermal development if permeability and fluids also exist. Song and Liu [37] identified 11 major geothermal resource regions based on anomalous geothermal gradients >35°C/km (**Figure 6**). These 11 high anomalous areas of geothermal gradient are the potential areas of geothermal development. Hot springs are prevalent through-out the 11 areas (**Figure 3**), indicating permeability and fluids. These resource areas occur either along the central range with the high rate of seismicity that sustained the 1999 M = 7.8 Chi-Chi earthquake (II, III, VII, **Figure 6**), along the Philippine fault within the Longitudinal Valley (IV, V, VI); both of which provide ample permeability for hot fluid circulation. Area I is within the Tatun volcanic area and permeability and fluid flow has been observed from exploratory wells [41, 53]. Area VII has high heat flow and hot springs, but no further confirmation of permeability has been identified. Taiwan has about 2500 mm of rain each year [38], sufficient for replenishing reservoirs.

In Taiwan, orogenic reservoirs are typically located at a shallow depth of 2–3 kilometers in the Central range and volcanic zone where the geothermal gradient is greater than 30° C/km. The development depth of enhanced geothermal away from the Central range is usually greater than 3000 m where the geothermal gradient is approximately 30 mW/M^2. Exploration is necessary to determine whether there is an organic reservoir or a candidate for an engineered system. Taiwan is currently conducting small-scale exploration to potentially exploit potential geothermal resources [39].

Author details

Chao-Shing Lee[1], Lawrence Hutchings[2*], Shou-Cheng Wang[1], Steve Jarpe[3], Sin-Yu Syu[1] and Kai Chen[1]

1 National Taiwan Ocean University, Keelung, Taiwan

2 Lawrence Berkeley National Laboratory, Berkeley, California, USA

3 Jarpe Data Solutions Inc., Prescott Valley, AZ, USA

*Address all correspondence to: ljhutchings@lbl.gov

IntechOpen

References

[1] Chen CH. Chemical characteristics of thermal waters in the central range of Taiwan, R.O.C. Chemical Geology. 1985;**49**(1):303-317. DOI: 10.1016/0009-2541(85)90163-9

[2] Chen CH. The hot spring and geothermal in Taiwan. Ti-Chih. 1989;**9**:327-340 (in Chinese)

[3] Sibuet J-C, Hsu S-K. How was Taiwan created? Tectonophysics. 2004;**379**:159-181

[4] Byrne T, Chan Y-C, Rau R-J, Lu C-Y, Lee Y-H, Wang Y-J. The arc–continent collision in Taiwan. In: Brown D, Ryan PD, editors. Arc-Continent Collision, Frontiers in Earth Sciences Series. Vol. 4. Berlin: Springer; 2011. pp. 213-245. DOI: 10.1007/978-3-540-88558-0_8

[5] Wu FT, Rau R-J, Salzberg D. Taiwan orogeny: Thinskinned or lithospheric collision? Tectonophysics. 1997;**274**:191-220

[6] Moore GF, Taira A, Bangs NL, Kuramoto S, Shipley TH, Alex CM, et al. Data report: Structural setting of the leg 190 Muroto transect, [CD-ROM]. Proceedings Ocean Drilling Program Initial Report. 2001;**190**:1-14

[7] Hilde TWC, Lee C-S. Origin and evolution of the west Philippine basin: A new interpretation. Tectonophysics. 1984;**102**:85-104

[8] Nagel S, Granjeon D, Willett S, Lin AT-S, Castelltort S. Stratigraphic modeling of the Western Taiwan foreland basin: Sediment flux from a growing mountain range and tectonic implications. Marine and Petroleum Geology. 2018;**96**:331-347. DOI: 10.1016/j.marpetgeo.2018.05.034

[9] Lester WR, Lavier L, McIntosh KD, Van Avendonk HJA, Wu F. Active extension in Taiwan's precollision zone: A new model of plate bending in continental crust. Geology. 2012;**40**(9): 2437-2440. DOI: 10.1130/G33142.1

[10] Huang H-H, Wu Y-M, Song X-D, Chang C-H, Lee S-J, Chang T-M, et al. Joint Vp and vs tomography of Taiwan: Implications for subduction-collision orogeny. Earth and Planetary Science Letters. 2014;**392**:177-191

[11] Rau RJ, Wu FT. Tomographic imaging of lithospheric structures under Taiwan. Earth and Planetary Science Letters. 1995;**133**:517-532

[12] Rau R-J, Ching K-E, Hu J-C, Lee J-C. Crustal deformation and block kinematics in transition from collision to subduction: GPS meurements in northern Taiwan, 1995-2005. Journal of Geophysical Research. 2008;**113**: B09404. DOI: 10.1029/2007JB005414

[13] Sibuet J-C, Zhao M, Wu J, Lee C-S. Geodynamic and kinematic context of South China Sea subduction during Okinawa trough opening and Taiwan orogeny. Tectonophysics. 2021;**817**:1-18. DOI: doi.org/10.1016/j.tecto.2021.229050

[14] Lee Y-H, Chen C-C, Liu T-K, Ho H-C, Lu H-Y, Lo W. Mountain building mechanisms in the southern central range of the Taiwan Orogenic Belt: From accretionary wedge deformation to arc-continental collision, earth planet. Science Letters. 2006;**252**:413-422. DOI: 10.1016/j.epsl.2006.09.047

[15] Ching K-E, Hsieh M-L, Johnson KM, Chen K-H, Rau R-J, Yang M. Modern vertical deformation rates and mountain building in Taiwan from precise leveling and continuous GPS observations, 2000-2008. Journal of Geophysics Research. 2011;**116**:b08406. DOI: 10.1029/2011jb008242

[16] Shan YH, Nie GJ, Ni YJ, Chang CP. Structural analysis of a newly emerged

accretionary prism along the Jinlun-Taimali coast, southeastern Taiwan: From subduction to arc-continent collision. Journal of Structural Geology. 2014;**66**:248-260. DOI: 10.1016/j.jsg.2014.06.002

[17] Ching K-E, Hsieh M-L, Johnson KM, Chen K-H, Rau R-J, Yang M. Modern vertical deformationrates and mountain building in Taiwan from precise leveling and continuous GPS observations, 2000-2008. Journal of Geophysical Research: Solid Earth. 2011;**116**:B8. DOI: 10.1029/2011JB008242

[18] Chang CP, Angelier J, Lu CY. Polyphase deformation in a new 5y emerged accretionary prism: Folding, faulting, and rotation in the southern Taiwan mountain range. Tectonophysics. 2007;**466**:436. DOI: 10.1016/j.tecto.2007.11.002

[19] Wu, F.T, C.F. Chang, and Y.M. Wu (2004) Precisely relocated hypocentres, focal mechanisms and active orogeny in Central Taiwan. From: Malpas, J., Fletcher, C. J. N., Ali, J. R. & Aitchison, J. C. (eds). Aspects of the Tectonic Evolution of China. Geological Society, London, Special Publications, 226, 333-354. 0305-8719/04/$15 9 The Geological Society of London 2004.

[20] Kao H, Chen W-P. The Chi-Chi earthquake sequence: Active, out-of-sequence thrust faulting in Taiwan. Science. 2000;**V**:288

[21] Ho C-S. A synthesis of the geologic evolution of Taiwan. Tectonophysics. 1986;**125**:1-16. DOI: 10.1016/0040-1951(86)90004-1

[22] Ramsey LA, Walker RT, Jackson J. Geomorphic constraints on the active tectonics of southern Taiwan. Geophysical Journal International. 2007;**170**:1357-1372

[23] Lin C-H. Thermal modeling of continental subduction and exhumation constrained by heat flow and seismicity in Taiwan. Tectonophysics. 2000;**324**(3):189-201. DOI: 10.1016/S0040-1951(00)00117-7

[24] Wu YM, Chang C-H, Zhao L, Shyu B, Chen Y-G, Sieh K, et al. Seismic tomography of Taiwan: Improved constraints from a dense network of strong motion stations. Journal of Geophysical Research. 2007;**112**:B08312. DOI: 10.1029/2007JB004983

[25] Chi W-C, Reed DL. Evolution of shallow, crustal thermal structure from subduction to collision: An example from Taiwan. Geology Society A, Bulletin. 2008;**120**(5/6):679-690. DOI: 10.1130/B26210.1

[26] Lin CH. Active continental subduction and crustal exhumation: The Taiwan orogeny. Terra Nova. 2002;**14**:281-287. DOI: 10.1046/j.1365-3121.2002.00421

[27] Burg JP, Gery TV. The role of viscous heating in Barrovian metamorphism of collisional orogens: Thermomechanical models and application to the Lepontine dome in the Central Alps. Journal of Metamorphic Geology. 2005;**2005**(23): 75-95. DOI: 10.1111/j.1525-1314.2005.00563.x

[28] Allemand & Lardeaux. Strain partitioning and metamorphism in a deformable orogenic wedge: Application to the alpine belt. Tectonophysics. 1997;**280**:157-169

[29] Beaumont C, Jamieson RA, Nguyen MH, Lee B. Himalayan tectonics explained by extrusion of a low-viscosity crustal channel coupled to focused surface denudation. Nature. 2001;**414**:738-742

[30] Burov E, Jolivet L, Le Pourhiet L, Poliakov A. A thermomechanical model of exhumation of high pressure (HP) and ultra-high pressure (UHP) metamorphic rocks in alpine-type collision belts. Tectonophysics. 2001;**342**:113-136

[31] Gerya TV, Meilick FI. Geodynamic regimes of subduction under an active margin: Effects of rheological weakening by fluids and melts. Journal of Metamorphic Geology. 2011;**29**(1):7-31

[32] Nabelek PI. Al partial melting and ultra high temperature metamorphism in convergent orogens: Implications of temperature-dependent thermal diffusivity and rheology. Journal of Geophysical Research;**115**:B12417. DOI: 10.1029/2010JB007727

[33] Lee C-R, Cheng W-T. Preliminary heat flow measurements in Taiwan. In: Fourth Circum-Pacific Energy and Mineral Resources Conference. Geological Society of Singapore. Singapore; 1986 August 17-22, 1986

[34] Barr TD, Dahlen FA. Brittle frictional mountain building, 2. Thermal structure and heat budget. Journal of Geophysical Research. 1989;**94**:436. DOI: 10.1029/88JB04332 Issn: 0148-0227

[35] Xia K, Xia S, Chen Z, Huang C. Geothermal characteristics of the South China Sea. In: Gupta ML, Yamano M, editors. Terrestrial Heat Flow and Geothermal Energy in Asia. New Delhi, India: Oxford & IBH; 1995. pp. 113-128

[36] Hwang WT, Wang C. Sequential thrusting model for mountain building: Constraints from geology and heat flow of Taiwan. Journal of Geophysical Research. 1993;**98**:9963-9973. DOI: 10.1029/92JB02861

[37] Song Shengrong and Liu Jiamei. Hot Springs in Taiwan. Annual rainfall reported. 2003. p. 208. Available from: https://www.cwb.gov.tw/eng/

[38] Yao M-H, Li M-H, Juan J-Y, Hsia Y-J, Lee P-H, Shen Y. Mapping reference evapotranspiration from meteorological satellite data and applications. Terrestrial, Atmospheric and Oceanic Sciences. 2017;**28**:501-515. DOI: 10.3319/TAO.2016.11.15.01

[39] Chao-Shing. Geothermal Resource Asssociation. National Ocean University Press; 2022

[40] Yeh YL, Wang WH, Weng S. Dense seismic arrays deny a massive magma chamber beneath the Taipei metropolis, Taiwan. Scientific Reports. 2021;**11**:1083. DOI: 10.1038/s41598-020-80051-4

[41] Dobson P, Gasperikova E, Spycher N, Lindsey N, Guo T, Chen W, et al. And fowler, a: Conceptual model of the Tatun geothermal system, Taiwan. Geothermics. 2018;**74**:273-297

[42] MIT. The Future of Geothermal Energy (Technical Report). Cambridge, MA (United States). Sponsoring Org.: USDOE Office of Energy Efficiency and Renewable Energy (EERE), Geothermal Technologies Office (EE-4G). OSTI Identifier: 1220063. DOE Contract Number: AC07-05ID14517: Massachusetts Inst. of Technology (MIT); 2006. p. 1083. DOI: 10.1038/s41598-020-80051-4

[43] Syu S-Y. S-wave attenuation in the geothermal reservoir and it's rock physics properties. National Taiwan Ocean University. Institute of Earth Sciences. Master's thesis Instructor: Li Zhaoxing. 2021:42

[44] Lin W, Tong L-T. Geological and geophysical surveys, preliminary report. In: Industrial Technology Research Institute (ITRI) Material and Chemical Research Laboratories (MCL). SNFD Team; 2020

[45] Chen K. The Geothermal Energy in the Metamorphic Belts in Taitung. East Taiwan: Thesis of National Taiwan Ocean University; 2021. p. 40

[46] Hutchings L, Bonner B, Saltiel S, Jarpe S, Nelson M. Rock physics interpretation of tomographic solutions for geothermal reservoir properties. Applied Geophysics with Case Studies on Environmental, Exploration and

Engineering Geophysics, AliIsmet
Kanlı, Intech Open. 2019:1-21.
Chapter 1. DOI: 10.5772/
intechopen.81226

[47] Gritto R, Hutchings LJ, Jarpe S,
Nihei KT, Schoenball M. Seismic
imaging of spatial- and temporal
heterogeneity of a geothermal reservoir
using a cost-effective dense seismic
network. In: Proceedings, 45th
Workshop on Geothermal Reservoir
Engineering. Stanford, California:
Stanford University; 2020 February
10-12, 2020, SGP-TR-215

[48] Hutchings L, Jarpe S, Boyle K,
Viegas G, Majer E. Inexpensive,
Automate micro-earthquake data
collection and processing system for
rapid, high-resolution reservoir
analysis. In: 2011 Annual Meeting. San
Diego, CA: Geothermal Resources
Council, Transactions; 2011

[49] Menke W. Geophysical Data
Analysis: Discrete Inverse Theory.
Academic Press, an imprint of
Elsevier; 2012

[50] Ernie M, James N, Ann Robertson T,
Jean S, Wong I. Us Geothermal
Technologies Office. January 2012.
Protocol for Addressing Induced
Seismicity Associated with Enhanced
Geothermal Systems.

[51] Induced Seismicity Protocol in
Energy Technologies. US National
Academy of Sciences. Committee on
Induced Seismicity Potential in Energy
Technologies. National Research Council.
2012. p. 300. ISBN: 978-0-309-25367-3

[52] Hutchings L, Bonner B, Jarpe S,
Singh A. Micro-Earthquake Analysis for
Reservoir Properties at the Prati-32
Injection Test, the Geysers, California. Las
Vegas, Nevada: Proceedings, Geothermal
Resources Council; 2014 2014

[53] Perdana P, Satria T, Chen B-C,
Hackett L, Robertson-Tait A, Thomas A.
Geothermal resource evaluation of the
Tatun volcano group (TVG) area,
Taiwan. In: PROCEEDINGS, 46th
Workshop on Geothermal Reservoir
Engineering. Stanford, California:
Stanford University; 2021 February
15-17, SGP-TR-218

Chapter 5

Geodynamics of Precambrian Rocks of Southwestern Nigeria

Cyril C. Okpoli, Michael A. Oladunjoye and
Emilio Herrero-Bervera

Abstract

The geodynamics of the Southwestern Nigeria Precambrian Basement Rocks were studied with aim of understanding the evolution of rocks globally. Magnetic carriers of Precambrian Basement rocks samples collected from 110 locations were prepared for rock magnetism, optical microscopy and Scanning Electron Microscopy (SEM). The Natural Remanent Magnetisation (NRM) of the remagnetised and unmagnetised rocks are strong (0.3–1.7 A/m -< 0.5 A/m) showed northwesterly direction with moderate inclination and weak NRM with westerly shallow direction respectively. Primary and secondary NRMs are carried by maghemite, and the remagnetised and unmagnetised rocks revealed a higher coercivity for alternating field demagnetisation (<20 mT – < 10 mT median destructive field). Optical microscopy revealed maghemite, poor titanomagnetite, titanomaghemite lamellae >30 pm and finer maghemite/magnetite grains finer than 10 pm. X-ray Diffratometry (XRD) and SEM results implied NW remanence in the remagnetised rocks reside in the fine poor-maghemite during the alteration of hornblende to actinolite while the coarse-grained maghemite in both rocks carries the W remanence of a thermoremanent magnetisation acquired in the Pan – African times. Global cold collision geodynamics resulted in the generation of ultra-high pressure metamorphic complexes and remagnetisation and True Polar Wander drifts of the paleomagnetic pole move towards the equator.

Keywords: remagnetised, unremagnetised, pan-African, tectonometamorphism, NRM, orogenesis

1. Introduction

Geodynamics of the Precambrian is a fascinating and contentious topic that is now preventing us from better understanding how the Earth evolved over time. The dearth of raw data related to this tectonic regime is largely to blame for the current controversy and lack of consensus on Precambrian geodynamics. Geodynamics is the study of how the interior and surface of the Earth change through time. A time-depth diagram (**Figure 1**) that spans the whole history and interior of the Earth can be used to show this process schematically. For a systematic characterisation of geodynamic interactions, data points characterising the physical-chemical condition of the Earth at different depths 0 to 6000 km, for discrete times in geological time, are shown in **Figure 1**. (ranging from 0 to around 4.5 billion years ago).

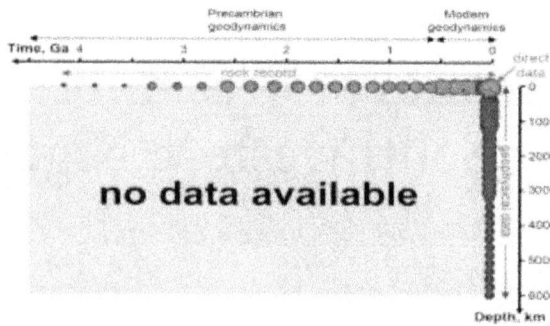

Figure 1.
Data availability for restricting the geodynamic connection for the earth is depicted in a simplified time–depth diagram. (modified after [1]).

Geophysical data measurements, unfortunately for geodynamics, provide systematic coverage of the current Earth structure and the geological record recorded in rocks formed near the Earth's surface (usually within a few tens of kilometres). As a result, Precambrian geodynamics remains a controversial topic. It's also worth mentioning that four key Precambrian Earth evolution topics are among the top ten questions defining 21st-century Earth sciences [2]:

1st "What happened during Earth's "dark period" of the first 500 million years?" This period is critical for understanding planetary history, particularly how the Earth's atmosphere and seas formed, yet scientists know little about it because few rocks from this age have been preserved."

2. "How did life begin?" Remaining records of geological examinations of rocks and minerals could be used to identify where, when, and in what and what form life first arose."

3. "How does the Earth's interior function, and how does it impact the surface?"

Earth's magnetic field was formed by the continual movement of the mantle and core.

How and when continents formed and were preserved throughout billions of years, as well as their future evolution.

In this study, we will concentrate on the last questions, which have improved dramatically over the last decade.

This progress has been fueled by an increase in the quality and quantity of geological, geochemical, petrological, and geochronological data for Precambrian rock complexes, as well as the ongoing development of analogue and numerical models for early Earth dynamics [3, 4]. Volcanism, seafloor building, and mountain formation are all aided by mantle convection, which has an impact on surface conditions.

Scientists, on the other hand, are unable to exactly characterise these motions or calculate how they differed in the past, making it impossible to comprehend the past and predict the Earth's future surface environment.

How did the Earth's plate tectonics and continents form?

Despite the fact that plate tectonic theory is widely accepted, scientists are still perplexed as to why Earth has plate tectonics and how closely it is tied to other planet features such as water content, continents, oceans, and life. Modelling has become increasingly important in generating new goods due to the shortage of empirical constraints (**Figure 1**).

Indeed, as Benn et al. [3] point out, one of the unique aspects of Precambrian geodynamics is that there is no thriving global geodynamics paradigm, and early Earth lithosphere tectonics differs from modern-day plate tectonics, which we can integrate and evaluate using our ever-growing set of observational and analytical data.

Several new major results have been obtained to address this particular challenge since Benn et al. [3]'s wide summary of Archean geodynamics, based primarily on merging geochemical, geological, petrological, and geophysical data sets. "This very concise, up-to-date synthesis of Precambrian geodynamics was motivated by analogue and numerical model results."

This research integrate modern paleomagnetic remanence, rock magnetism and optical microscopy and concepts (Plate tectonics and subduction, Orogeny and collision), petrology (metamorphic parageneses and relation with deformation), and geochronology. Data from regional literature including geophysical data are also abundantly used for synthesis and re-interpretation. The most important achievements include: the paleomagnetic studies and geodynamics on plate tectonics and subduction and orogeny and collision.

2. Regional settings of Precambrian geodynamics

The southwestern Nigeria granitoids is within the basement complex domain that was reopened in the Pan- African time of the Neoproterozoic period. This province was located around East Saharan, southeast Congo craton and west of West African craton (**Figure 2**), and has a long stretch from Hoggar to Brazil, which ranges from 4000 km to an extensive orogen in hundreds of kilometres [7]. The Trans-Saharan fold belt runs north-southerly, and the reopening of this belt was due to East Saharan, Congo and West African cratons continental collision about 790 and 500 Ma [6, 8, 9]. Granitoids, growth of thrust-nappe, medium- to high-grade metamorphism, parallel orogen tectonics typifies this belt [10]. The Hoggar separated into Air, Eastern and Central polycyclic; but now called the Pharusian belt plus Laouni terrain Algeria (LATEA) microcontinent [11]. Aggregation of twenty-three micro terranes constitute eastern and polycyclic central Hoggar in the northern province [5], while in the southern block (Dahomeyide), we have the Aïr-Hoggar composed of various continental oblique collisions [12].

Nigerian sector evolved by profuse magmatism in late Neoproterozoic times at the culmination of prior basin made up of depleted Archaean crust [13]. The Nigerian part of the Dahomeyide was separated into the eastern (granulite

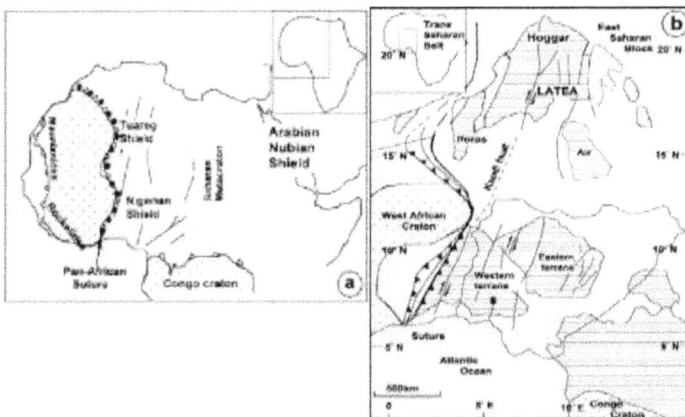

Figure 2.
Regional geological map of trans-Saharan metacraton/shield (modified after [5, 6]).

facies) and western (greenschist to amphibolites facies) domains based on some petrological attributes [14]. The southwestern Precambrian granitoids consist of migmatite-gneisses, schists, granites and dykes [15]. Pan-African granitoids and unmetamorphosed dykes are assigned Neoproterozoic isochron [8, 16–20]. The Archean crust characterised the supercrustals, which later interpreted to be deposited in diverse proto ocean floors [12, 15].

Pan-African belt evolution was by Plate tectonism, which led to the active margin colliding with the Pharusian belt and passive continental margin of the West-African craton about 600 Ma [7, 21–23]. The existence of basic to ultra-basic rocks thought to be remnants of mantle diapers or paleo-oceanic crust is part of this fact, and they have complex ophiolitic characteristics. Geochronological studies have examined major magmatic complexes with their isochron ages varying from 557 ± 8 to 686 ± 17 Ma (Rb/Sr. whole rock); 640 ± 15 Ma (U–Pb), which were determined in these complexes. Deformation of migmatite-gneisses and post-tectonic uplift typifies the Pan-African fold belt in southwestern Nigeria; which consist of polycyclic orogen of Liberian (2700 ± 200 Ma), Eburnean (2000 ± 200 Ma), Kibaran (1100 ± 200 Ma), and Pan-African (600 ± 150 Ma) [7, 24]. For the Liberian and Eburnean, the International Geological Time Scale (2002) has followed the following ages: "Paleoarchean to Mesoproterozoic (3600 to 1600 Ma)", "Mesoproterozoic to Neoproterozoic (1600 to 1000 Ma)", "Neoproterozoic to Early Paleozoic (1000 to 545 Ma)", and "Neoproterozoic to Early Palaeozoic (1000 to 545 Ma [25].

Ferre *et al.* [6] studied the northeastern Nigeria Pan-African continental collision, which resulted in high grade - high pressure and temperature (HP-HT) metamorphism up to granulite facies, migmatisation in supracrustal units of the same tectonism as the southeastern Nigeria domain [26, 27]. The extensive Archean crust of northern Nigeria was modified and remelted during the Pan-African tectonometamorphic episode. The Pan-African nappe system rejuvenates older polyorogenic times [15, 28].

Precambrian basement rocks into four units: Migmatite-Gneiss (migmatites, gneisses, granite-gneisses); Schist zones (schists, phyllites, pelites, quartzites, marbles, amphibolites); Pan African Granitoids (granites, charnockite, granodiorites, diorite, monzonites, gabbro) and Undeformed Acid and Basic dykes (muscovite, tourmaline, pegmatites, aplites, syenites, basaltic, dolerites and lamprophyre dykes). They occur as a small medium-grained rock with massive hills. This charnockites is made up of orthopyroxene, clinopyroxene, hornblende, plagioclase, alkali feldspar, magnetite, quartz and zircon. In some places, granite, porphyritic, augen gneiss, banded gneiss can be seen as low-lying outcrops and large hills (**Figure 3**).

2.1 Gondwana configuration

Although different models exist for the absolute position of Gondwana [30] as well as the relative positioning of cratons can be done with small margins of error (**Figure 4**; [32]). The formation of Gondwana is often presented as a merger of East Gondwana (Antarctica, Australia, and India) with West Gondwana (those currently in Africa and South America). However, evidence, especially from the eastern Gondwana cratons, indicates that it was not a simple unification of two halves but rather a poly-phase amalgamation of cratons during the waning stages of the Proterozoic, as a result, that Gondwana was created [33].

The Congo and West Africa cratons form part of West Gondwana and are connected through the Borborema Province in northern Brazil (**Figure 4**). This province was essentially an assemblage of several terrains and comprised reworked Mesoproterozoic- Neoproterozoic metasedimentary rocks and

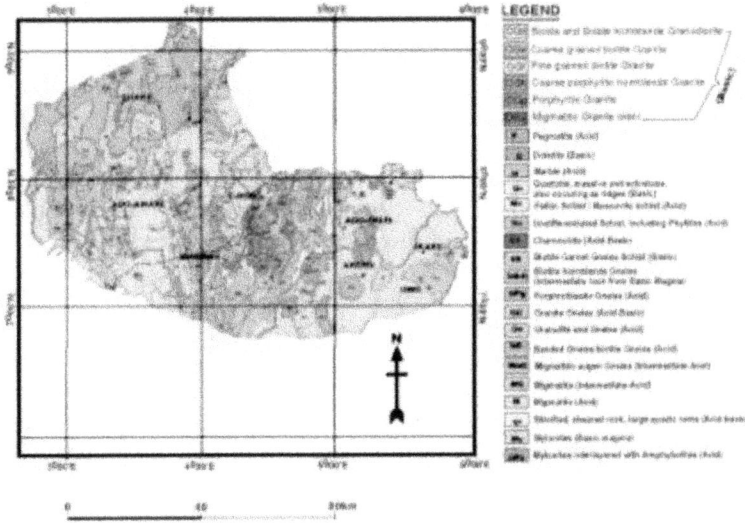

Figure 3.
Geological map of southwestern Nigeria (modified from NGSA, 2010 [29]).

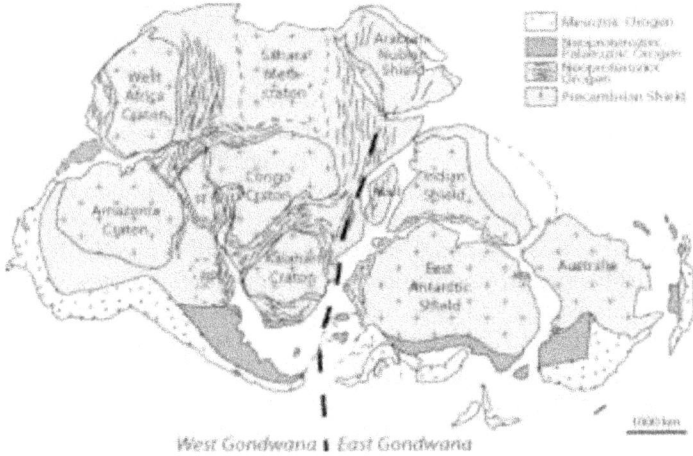

Figure 4
Gondwana map with its cratonic nuclei positions (adapted after [31]). RP -Rio de la Plata craton; SF- Sao Francisco craton.

Archean-Palaeoproterozoic crystalline basement [34]. Reworking is the result of Neoproterozoic continent-continent collision, which caused extensive deformation, migmatisation, granitisation and intrusive plutons. Geochronological constraints for the different stages of deformation in the Borborema Province are provided by U-Pb radiometric ages of the granitoid plutons [35]. Ages for zircons from syn-tectonic I-type granitoids and zircons from migmatitic gneisses show that deformation started ca. 625 Ma and peaked at about 600 Ma [36]. Post-tectonic alkaline granitoids mark the final orogenic stage, and U-Pb zircon ages show that deformation had ceased around 570 Ma [36]. The Borborema Domain was correlated, predominantly based on Sm-Nd model ages and U-Pb zircon ages of Archaean-Palaeoproterozoic basement rocks in conjunction with Neoproterozoic structural tectonic data, with the

Central African fold belt (**Figure 4**) and with the Nigerian Shield (**Figure 4**) in NW Africa [35].

The Central African fold belt demonstrate a poly-stage geodynamic evolution of nappe emplacement onto the Congo Craton northwardly [37]. Geochronological constraints reveal a history of individual orogenic stages broadly coeval with those of the Borborema Province: high-pressure metamorphism with granulite facies typified for syn-tectonic calc-alkaline and S-type granitoids and migmatisation occurred at 640–610 Ma, as well as post-collisional phase of exhumation and late-tectonic calc-alkaline to sub-alkaline granitoid emplacement was dated at 610–570 Ma [37]. The exact nature of the continental landmasses involved was still enigmatic. The belt could be entirely the consequence of the collision of the Congo Craton with the ill-defined Saharan Metacraton [38].

Neoproterozoic intrusions within the Nigerian Shield show a history very similar to that of the Borborema Province [34]. Combined structural data and U-Pb ages suggest that an early deformational phase took place at 640–620 Ma, peak meta-morphism and syn-tectonic granitoids are positioned between 620 and 600 Ma and a post tectonic phase from 600 to 580 Ma (geochronological data synthesised by [34]). Geochronology of these plutons shows that the continental collision evolved diachronously between 620 and 580 Ma [5]. The Nigerian Shield and the Tuareg Shield were parted from the West African Craton by the Dahomeyide and Pharusian belts (**Figure 4**). Peak metamorphism in the Dahomeyides occurred approximately at 610 ± 2 Ma [39] to 603 ± 5 Ma [40] premised based on radiometric dating of U-Pb obtained from gneisses from granulite-facies peak metamorphic zones. The post-collisional exhumation was dated by 40Ar-39Ar muscovite ages of 587 ± 4.3 and 581.9 ± 2.4 Ma [39], which corresponds with rutile ages of 576 ± 2 Ma, which repre-sent regional cooling below 400°C [40]. The collision of Island Arc with the West African Craton around 620 and 580 Ma, simultaneous with the height of the tectonic events in the Tuareg Shield to the east [5]. The Borborema domain evolved synchro-nous with the Central African Fold Belt, the Nigerian Shield (Dahomeyide Belt), and the Tuareg Shield (Pharusian Belt) strongly implies that this part of West Gondwana had amalgamated by 600 Ma, and all tectonic activity had ceased by 570 Ma.

Archaean to Mesoproterozoic granite-gneiss-migmatite complexes, greenstone belts and metasedimentary and metavolcanic units are caught up in the Brasilia Belt involving the Sao Francisco Craton and Magmatic Arc of Goias [41]. Observed data from the Paraguay-Araguaia belt that flanks the Goias Arc's western side imply that the collision of Sao Francisco/Goias with Amazonia slightly post-dates the Brasilia event at ca 550 Ma [42, 43]. Biotite and muscovite ages around 530 Ma from Archaean basement gneisses may record late-orogenic cooling in the Araguaia belt (K-Ar) [43].

3. Materials and methods

We make use of the following instrument to conduct measurements on some Precambrian basement rocks of southwestern Nigeria:

1. **Alternating frequency, thermal demagnetisations and Spinner magnetometer for Paleomagnetic analysis**.

Granite biotite granite gneiss, banded gneiss, Augen gneiss, porphyritic granite, syenite were selected based on their mineralogy, magnetic susceptibility, and natural remanent magnetisation (NRM). A combination of alternating field (AF) and thermal (TH) demagnetisation methods were employed. Their primary and secondary multi-remanence constituents were measured using the equipment.

These techniques were used because constituent minerals obtained through different mechanisms have different coercivity spectra and blocking temperatures. The coercivities of magnetic minerals are involved in AF demagnetisation. The alternating field method entails exposing the specimen to increasing amounts of AF, with the waveform being sinusoidal and decreasing in magnitude linearly with time. It was used to extract remanence from grains whose coercivities were less than the peak demagnetising area. The alternating magnetic field is a quick treatment procedure likened to the thermal demagnetisation method. Test of the natural remanent magnetisation in determining the rock material is not a superimposition of several magnetic constituents, and this was done by isolating the components of stable magnetisation (CRM).

The specimens were heated to a temperature below and near ferromagnetic mineral Curie temperatures in steps of 30 and 50°C during step-by-step thermal (TH) demagnetisation and then cool in a zero magnetic field at room temperature. It gives magnetic grains blocking temperatures (Tb) lower than the temperature used to strip a portion of their normal remanent magnetisation. Step by step, temperature ranges were measured, and residual magnetisation and susceptibility were calculated. The basic measurement of NRM yields the remanent magnetisation recorded in rocks (declination, inclination, and total intensity). In the present study, the samples were AF demagnetized in 14 steps following a sequence 2.5, 5, 7.5 10, 12.5, 15, 17.5, 20, 25, 30, 40, 60, 80 and 100 mT respectively. The thermal demagnetisation was done on some selected samples in a sequence of 50°C, 100°C, 150°C, 200°C, 250°C, 300°C, 350°C, 400°C, 420°C, 440°C, 460°C, 500°C, 530°C, 550°C and 570°C respectively.

As soon as alternating field and thermal demagnetisation are treated, the rock specimen directions are studied to isolate magnetic constituents. In this paleomagnetic study, stereographic and orthogonal projections were adopted. Stereographic projection direction was characterised, magnetisation vector unit length tip was measured, the same sphere diameter aligned with the southern pole. They are the contact site with the equator plane sphere, usually referred to as a small open circle. The geographic directions of the north, east, south and west were defined. Magnetic declination ranges from 0° (N direction) to 360° clockwise, and from 0° at the edge of the equator plane to 90° at the midpoint. AF and thermal datasets were analysed with AGICO's Remasoft 3.0 program [44] and Demagnetisation analysis in excel DAIE-v2015 program [45]. Fisher [46] 's statistics were employed to measure mean orientations.

2. Kaiser 785 nm Micro-Raman Microprobe system and Renishaw 830 nm inVia micro-Raman to determine iron oxides.

Granite; biotite granite gneiss; Augen gneiss and banded; banded gneiss; porphyritic granite; syenite, and amphibolite rock samples were hammered bits and pieces and selected with a solid permanent magnet in the laboratory because of their mineralogy and magnetic susceptibility in order to determine the magmatic effect of maghemite. Tiny, unpolished grains of different iron titanium oxides concentrations were affixed on carbon tape attached to a glass slide for Raman spectra measurements. In addition to optical images, micro-Raman spectroscopy of various excitation wavelengths was used. At the University of Hawai'i, various instruments were used to capture Raman spectra. Spectra with 785 and 830 nm - Kaiser Optical Systems' micro-Raman system and Renishaw in Via microspectroscopy were used for the study. The system consists of a 785 and 830 nm Invictus diode laser, a Kaiser Holospec/Renishaw spectrometer, a spectral range of 150–3300 cm^{-1}, a Leica microscope with imaging capabilities, as well as an Andor CCD camera. The laser light and Raman pulse are sent to the microscope and spectrometer via a 100-meter optical fibre. A 50 objective lens fixed on the microscope in backscattering geometry was used to focus the laser spot and observe the signal. The spectrometer and microscope are fixed through optical mirrors of different wavelengths and were operated using a PRIOR workstation (via WiRE 3.2 software). The spectra were imported into MATLAB 7.4.0 and Grams/AI v8.0 for normalising statistical analysis, background interference in each spectrum, as well as baseline diffraction patterns, i.e. correction and peak fitting using Gaussian and Lorentzian geometries. Background correction was done using sixth-order polynomials in both cases. Principal component analysis (PCA) and significant factor analysis (SFA) were employed to determine the principal components. Specimen were stored to avoid artefacts, and laser power had below 0.7 mW to prevent the destruction of the specimen; neutral density filters had a constant power of 675 µW; acquisition time was 60 s; spectrometer calibration before acquiring Raman spectra; and cyclohexane standard protocols were used [47, 48].

3. **Scanning electron microscopy (SEM) (JEOLJSM-5900LV) and X-ray diffractometry (XRD) for magnetic mineralogy.**

Gneiss, granite, biotite-granite-gneiss, charnockite, and granite were used to describe the ferromagnetic minerals based on their mineralogy and magnetic susceptibility. The thin polished sections were studied using SEM. Thus, SEM and XRD were employed in the Institute of SOEST-HIGP (Manoa, Hawaii, USA) to constrain the mineralogy of accessory minerals. SEM was used for imaging, qualitative analysis (equipped with an "EDS") and quantitative analysis (when equipped with an "EDS/WDS"). SEM and EMPA were applied to characterise the specimen for Mineral identification; compositional information, microstructures/deformation and compositional evolution of minerals.

4. We modelled the tectonometamorphism episodes in the Precambrian era to picture the evolution of the Precambrian rocks of southwestern Nigeria and relate them to present-day orogenesis.

4. Results and discussions

4.1 Paleomagnetic results

For samples of the same site, CO-23c subjected to thermal treatments have secondary remagnetisation averagely 70% at 300–500°C the remainder of the signal was washed up to 570°C (**Figure 5d**). **Figure 5**(d-i) demonstrated weak magnetic coercivity, unstable remanent directions and abrupt changes in intensity in Zijderveld curve plotting not directly to the origin due to tectono-metamorphic episodes. Up to 500°C, the rest of the samples retains >50% of the magnetisation is lost and it It was difficult to isolate the ChRMs (e.g., CO-23c, CO-018,CO-37a, CO-74 N and CO-100A in **Figure 5a-e**). The unblocking temperature revealed two distinct elements, one with natural polarity against N and the other towards NNW and NE. The second specimen has a low unblocking temperature and was thoroughly cleaned up to 300°C, while the northerly specimen reported magnetisation up to intermediate unblocking temperatures (580°C), which is referred to as characteristic remanent magnetisation (*CRM*). Thus, regardless of NRM decrease for the first 300 samples, a large percentage of the samples were treated to remove secondary remanence [49]. Compared to the ferrimagnetic one, the samples have a low paramagnetic effect, as shown by the similarities of the curve before and

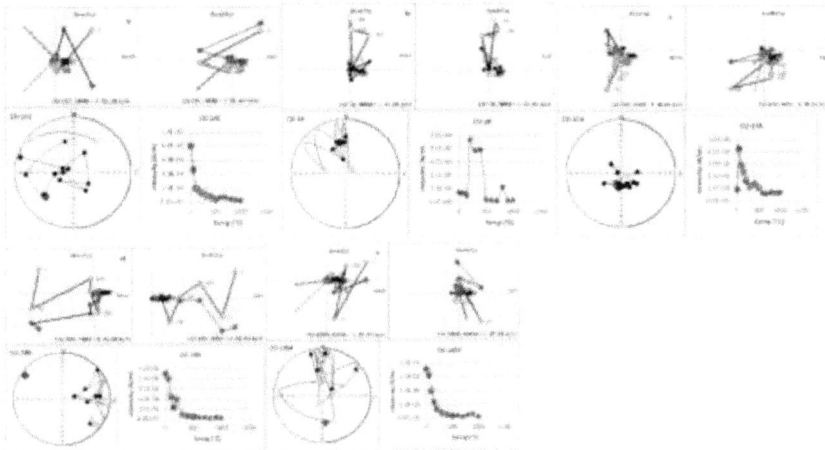

Figure 5.
(a-ce) alternate field and thermal demagnetization showing the Zijderveld orthogonal vector diagram of unremagnetised Precambrian rocks and (d-c) remagnetised rocks.

after slope correctionCO-23c, CO-018,CO-37a, CO-74 N and CO-100A samples were decomposed into two overlapping modules with medium destructive fields (MDF) ranging from 30 to 40 to 60–70 mT, and a third higher coercivity segment (*MDF* ~ 467 mT).

Determining the time of growth during folding is difficult because syntilting results observed in incremental tilt tests do not give a unique result. The formations of new minerals (maghemites) were demonstrated in most of the granitoids invoked for many syntilting CRMs. The remanence-carrying Fe oxide grains may have rotate during folding and tectonometamorphic episodes as a result of syntilting, which would alter the original magnetic direction. The rotated directions were not related to the ambient field during folding. Shear strain during flexural flow folding could cause a prefolding magnetization to be rotated into a syntilting configuration. Folds with different geometries and tilted thrust sheets all have the same magnetic characteristics and are probably caused by the same remagnetization events. The determined tilt test results however suggest that the CRM is pre-tilting in both the thrust sheets and a fold with a fault-bend fold geometry and syntilting in folds with a fault propagation fold geometry that probably experienced higher strains. A primary remanent magnetisation should theoretically require more stress than most rocks have been subjected to during deformation, be partially reversible and have the greatest effect on the low-coercivity. Developing a better understanding of remagnetization processes and use of palaeomagnetism for its studies, the preponderance of multi-domain and pseudo-Single domain magnetic phases and presence of maghemite suggest that the type 1 magnetite has been modified during the orogenesis. They are correlated to their respective bedding tilt orientation base on correlation fold test. The tectonic correction brought the site direction in its geographic coordinates; the same rotation was applied to the mean geographic coordinate to produce forward corrections of the mean. The remagnetised component of the granitoids was interpreted as partial thermoremanentmagnetisation (pTRM) overprint acquired during the tectonometamorphic episodes 600 ± 150 Ma (Pan –African orogeny) associated with tectonic accretion along southwestern Nigeria Precambrian shield. The overprints directions were likening to the reverse magnetization. The intermediate degree of unfolding at peak concentration could be due to subtle amounts of component mixing, diachronous magnetization acquired during a short time interval, syn-folding magnetisation acquisition, local structural differences within the fold.

Precambrian rocks of southwestern Nigeria with an intermediate unblocking temperature of 100—400°C witnessed perfect dispersed clustering in geographic coordinates after tilt correction, suggested remanence imprint after folding (**Figure 5a-c**).). The corresponding imprint on the paleomagnetic pole situated at 85.1°N, 183.0°E with α95 = 10.1° (dp = 12.7, dm = 8.0) in geographic coordinates, are very similar to the Basement system of Precambrian poles of southwestern Nigeria. The majority of the sites were unable to isolate the intermediate temperature component due to its remagnetisation ((**Figure 5d**). Therefore, this overprint was considered to be remagnetisation in Pan-African times. In Nigeria's Eastern basement complex, NE Brazil, Central Cameroon, and much of the west Gondwana crystals provinces, different remagnetisation has been observed.

Available geologic model isochron ages, tectono-metamorphic history, and crustal evolution model that support accretional model for Paleoproterozoic and Neoproterozoic rocks with consistent older model isochron age in support of significant involvement of Archean felsic crust in their orogeny and suggested that southwestern Nigeria's tectono-metamorphic history and crustal Nigerian active margins and Trans Saharan belt utilised U–Pb geochronological to infer the magmatism that occurred from 670 to 545 Ma ([26] and this study) for the overriding plate of Benino-Nigerian Shield. The age of ultra-high pressure metamorphic eclogites from the passive margin of the West African Craton subducted to mantle depths recently restricted the timing of crustal deformation to 600–150 Ma. The lower plate (West African Craton) and upper plate (Benino-Nigerian Shield) both experienced east-plunging continental subduction, which pushed the passive flanking margin to >90°, which suggested that granitoids subducted between 670 and 610 Ma. Both plates crystallised the Pharusian oceanic plate, while igneous rocks associated with arc magmatism include the hornblende-biotite granodioritic gneiss, dated at 610–694 Ma. Pan-African granites older than 610 Ma predate non-subduction-zone deformation. The Benino-Nigerian basement complex was formed by continental arc, according to a geochemical dataset and Sr-Nd isotopic kinematics application to the 670–610 Ma migmatite-gneisses.

Results exhibited preliminary records of exsolved maghemite in silicate, plagioclase and pyroxene minerals in the southwestern Nigeria Precambrian gneiss and granitoids. These iron oxides are seen in the pyroxene and plagioclase minerals showing good magnetic stability (**Figure 6a**). **Figure 6a** (i-iii) shows the site mean directions of the study area. Site mean direction for component i clustered around mean Dm = 325.6° lm = 28.4°(N = 12, α95 = 9.8, k = 10.93), which resulted to paleomagnetic pole located at 7.38°N. 5.57°E (A95 = 9.8, K = 12.9). Site mean direction for

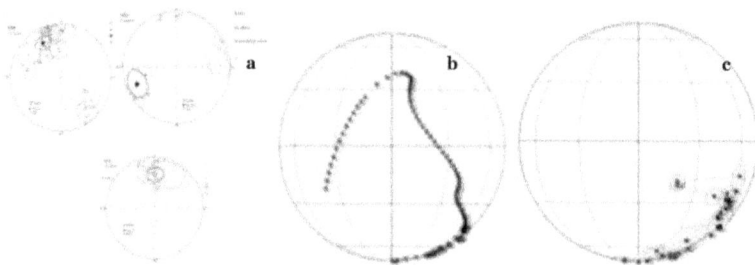

Figure 6.
(a) Site mean directions of the study area (b) spline apparent pole wandering APW of Africa pole and (c) VGP reconstruction of Africa databasefit (lat = 0.0, long = 151.60, angle = 27.50) with Africa, taking account of the opening of the South Atlantic.

component ii Dm = 0.6° lm = −38.4° ((N = 7, α_{95} = 14.1, k = 15.24), which resulted to paleomagnetic pole located at 7.24°, N 5°E (A_{95} = 12.5, K = 13.4). Site mean direction for component iii is Dm = 225.8° lm = 26.4° (N = 4, α_{95} = 16, k = 17.93), which resulted to the paleomagnetic pole located at 8.17.5°N 4.21°E (A_{95} = 15, K = 13.9). Thus, site mean directions of the study area were accomplished as a reliable recorder of the old geomagnetic fields. **Figure 6b-c** were employed to demonstrate the approximate tilt estimates, when distinct geological criteria demonstrated folding and faulting cannot be correlated to the unit. They used to evaluate the pole position for the SW Nigerian shield, errors due to (1) the large elliptical confidence was defined for the pole position, (2) recognised tilt events, and (3) the streaked great circle distribution of reversed, normal and mixed Remanent magnetization (RM) directions towards the Pan African events must be examined. Because the pole was poorly constrained, the data can accommodate several possible interpretations. First, the unit may have gained an RM orientation during long-term cooling compared to steady African cooling, and magnetic acquisition in the gneiss may have taken place over long stretches of time, during which reversals in the Earth's magnetic field direction occurred. Both polarities may have been recorded in individual sample owed to varying blocking temperatures. Perhaps, the field direction changed slightly during this period, then a deflection along a great circle towards the polarity which the earth's field sustained for the shortest length of time, would be observed. Such a deflection, where changes in the field direction are represented by angular limits of mean directions from Van Der Voo, [50] and percentages represent the relative time the earth's field spent in each polarity. These results in conjunction with demagnetization data suggest that the geomagnetic field was represented by both polarities (reversed dominating) during the magnetization of the southwestern Nigeria granitoids. The evident polar wandering path of Precambrian rocks in southwest Nigeria was caused by the effects of the mantle and superplumes. The acquired conjugate poles lie towards SW-NE at 304.8°E and 61.8°S directions (dp = 5.4, dm = 10.7); which was relatively at mean direction of 305.1°E and 64.5°S (dp = 2.3, dm = 4.5).

The Precambrian Basement rocks of southwestern Nigeria demonstrated low coercivities and low unblocking temperatures especially in the coarser grains.. No younger geological event occurred in the study area. The mafic syenite dykes witnessed in the study area occurred during the last phase of deformation. The tectonometamorphic history is interpreted as remagnetisation due to Neoproterozoic Pan-African event which is correlated to Brasilian 650 ± 150 Ma. The remagnetisation was not directly attributed to tectonic stress but to fluid, chemical and viscous effects, The remagnetisation phenomenon is due to new mineral growth, whose chemical remanent magnetization (CRM) swamps that of magnetically softer earlier grains. The progressive demagnetization, along the great circle path from initial towards pressure-vessel field. Only the initial remanence and the demagnetization fields determine the remagnetisation paths

4.2 Tectonic implications

The area has been tectonically moved as a result of post-magnetization causing the Nigerian shield magnetic direction to be deviated from an original primary direction. At least three explanations for the observed SW Nigeria paleopole position exist, these are: present position acquired magnetic remanence; NW thrust to after RM acquisition brought the SW craton to its present position, or the southwestern Nigeria acquired its magnetization and was transported to the northeast >1000 km by a left-lateral transcurrent fault system, and was later thrust to the northwest to the present location of the body. The magnetic fabric exhibited by the SW Nigeria granitoids closely approximates the mineral fabric, suggesting that both were acquired during the four deformational (D1, D2, D3 and D4) events which have affected the body. Results indicate that the RM was acquired during magnetite recrystallization or cooling from metamorphic temperatures (~600°C) to maghemite. RM acquisition in the gneiss seems to have happened over a comparatively long period of time, with reversed, mixed and normal polarities represented in the magnetic signature of the unit.

The introduction of hydrothermal fluids occurred at a temperature well above the Pb–Pb closure, which corresponds to the age of the magnetic pole in south-western Nigeria (~ 571 Ma). Granitoids emplaced over 700 Ma were not reliant on high-level hydrothermal emplacement in unmetamorphosed southwestern Nigerian Precambrian rocks, implying a Pan-African episode. Thus, the Pb–Pb date, which is younger than the 620 Ma U–Pb acquired from a deformed syenite, can provide the thrust a better constrain age. As a result, 620 Ma was proposed as the magmatic rock's crystallisation period rather than the tectonic event's age. The 571 Ma periods, on the other hand, proposed retrograde metamorphism at the nappe's base. The metamorphism ranges from amphibolite-granulite and retrograde greenschist facies towards north to base of nappe respectively in Nigeria.

5. Discusssion on Precambrian geodynamics in relation to paleomagnetism

5.1 Plate tectonics and subduction

It's worth mentioning that thermomechanical numerical experiments were only recently employed to study the onset and patterns of Precambrian plate tectonics and subduction [51–53]. Van Thienen et al. [54] presented one of the most well-known attempts to employ numerical modelling to analyse global tectonic processes of the early Earth).Since of the considerably different temperature and viscosity conditions that existed beneath the early Earth's mantle, and because current-day geodynamics cannot be easily projected back to the Earth's early history. Van Thienen et al. [54] used computational thermochemical convection models with partial melting and a basic mechanism for melt segregation and oceanic crust formation to investigate an alternative set of dynamics that may have been active in the early Earth. They are: Small scale convection involving the lower crust and shallow upper mantle; Large-scale resurfacing processes in which the entire crust sinks into the (ultimately lower) mantle, forming a stable reservoir of incompatible elements in the deep mantle and segregating melt builds a fresh crust on the surface and excessive melting and crustal growth due to the intrusion of lower mantle diapirs into the upper mantle at a high excess temperature (about 250 K). This allows plumes in the Archean upper mantle to have substantially higher excess temperatures than previously thought possible based on theoretical considerations. Various geodynamical theories have predicted a dense enriched layer at the mantle's base [55, 56]. Massive scale sinking of the thick basaltic/eclogitic crust induced by decompression melting of mantle peridotite, according to Van Thienen et al. [54], may have developed such a layer over a brief period early in the Earth's mantle's history. The large-scale crustal sinking model described by Van Thienen et al. [54] might thus be considered an alternative (or predecessor) to Albarede & Van der Hilst's proposed subduction model (2002).Large-scale sinking appear like subduction and could be a Precambrian forerunner to modern plate tectonics.

In contrast to other studies, O'Neill et al. [57] presented an alternate explanation for crustal growth episodicity by providing paleomagnetic evidence for periods of rapid plate motions matching to observed peaks in crustal age distribution. The Nd and Sr. isotope ratios of many juvenile terrains [58] support the idea of increased plume activity associated with these overturns and provide a model for their cra-tonization. Superplumes and other ideas have been presented to explain this episod-icity [59]. Plate-driven episodicity arises naturally in response to the early Earth's high mantle temperature, and hence can explain quick pulses of plate motion and crustal formation without the need for mantle overturn events [57]. To assess the possibility of subduction in the hotter Precambrian Earth, Van Hunan and van den Berg [53] employed 2D thermomechanical models with a single subduction zone enforced by a weak fault (**Figure 7**).

In contrast to Davies [61], instead of focusing on changes in crustal thickness spanning 10 to 22 km, this model ignores early upper mantle depletion; caused by rising mantle temperatures, that decided plate tectonics' viability on a hotter Earth. Numerical results revealed no Ultrahigh-Pressure Metamorphisms (UHPM) or blueschists in most of the Precambrian: early slabs were too weak to provide a mechanism for UHPM and exhumation. Due to the lower viscosity and higher degree of melting, a hotter, fertile mantle would have resulted in a thicker crust and a thicker depleted harzburgite layer in the oceanic lithosphere, according to van Hunan and van den Berg [53]. A thicker lithosphere may have been a significant stumbling block to subduction, and Earth in the Precambrian may have been characterised by a dif-ferent mode of downwelling [62] or "sub-lithospheric" subduction [53], though the conversion of basalt to eclogite may greatly relax this limitation [54, 63]. The natural reduced viscosity of the oceanic lithosphere on a hotter Earth would lead to increased Slab breakoff (**Figure 7**) and crustal detachment from the mantle lithosphere has occurred in some situations. Hence, lithospheric weakness may limit the feasibility of present plate tectonics on a hotter Earth. By merging knowledge from geochemical data and numerical models, Halla et al. [51] used inferences from van Hunan and van den Berg's [53] numerical study to constrain early Neoarchean (2.8–2.7 Ga) plate tec-tonics. Sizova et al. [60] employed a two-dimensional (two-dimensional) petrologi-cal–morphological model. To investigate the dependence of tectonic-metamorphic and magmatic regimes at an active plate margin on upper-mantle temperature, crustal radiogenic heat production, and lithospheric weakening, a thermomechanical numerical model of oceanic–continental subduction (**Figure 7**) was used to conduct

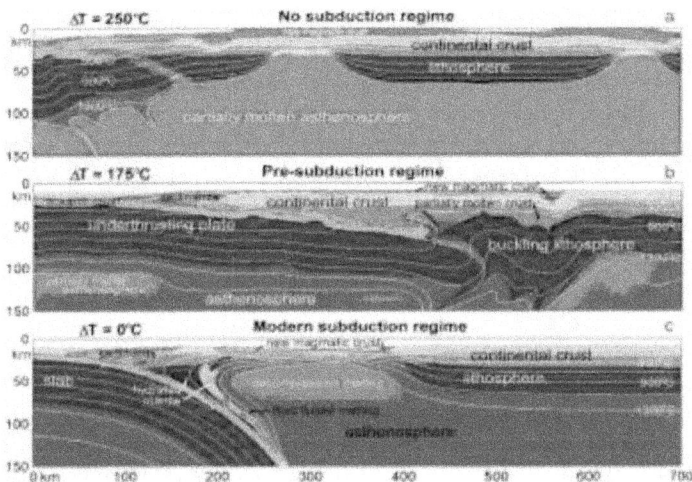

Figure 7.
Modelling the evolution of an active continental margin with high-resolution numerical models for various mantle temperature differences (T) above current values (modified after [60]).

a series of high-resolution experiments. Based on their testing, the scientists observed a first-order change from a "no-subduction" tectonic regime to a "pre-subduction" tectonic regime, and then to the current mode of subduction (**Figure 7**). The first transition is gradual and occurs between 250 and 200°C over current upper-mantle temperatures, whereas the second transition is abrupt and occurs between 1 and 2°C above current upper-mantle temperatures. The change to the current plate tectonic regime occurred at 3.2–2.5 Ga, according to the link between geological evidence and model results. Convergence does not result in self-sustaining one-sided subduction in the "pre-subduction" tectonic regime (upper-mantle temperature 175–250°C above the surface), but rather two-sided lithospheric downwellings and shallow under-thrusting of the oceanic plate beneath the continental plate (**Figure 7b**).

5.2 Orogeny and collision

Interpretations of geological, petrological, and geochemical data from Proterozoic and Archean orogenic belts revealed that the Precambrian had different tectonic kinds of orogeny than the present-day Earth [64–66]. Accretionary and collisional orogens are the two forms of Precambrian orogens [66–68]. When the oceanic crust is subducted along active continental margins, accretionary orogens form ([66] and references therein). Precambrian accretionary orogens make a significant contribution to continental expansion compared to Phanerozoic accretionary orogens due to their high rates of juvenile crust growth [66]. Several post-Archean accretionary orogens are terminated by continent-continent collisions during supercontinent formation [66]. The average terrain lifespan during the Archean is 70–700 million years, 50–100 million years during the pre-1 Ga Proterozoic, and 100–200 million years in later orogens [66]. When continents collide, collisional orogens emerge; they initially arose in the Proterozoic, but had little impact on continental growth [67, 68]. The appropriateness of studies of current collisional orogens to the Precambrian is yet unknown, given the impact of a warmer continental crust and a higher mantle on the geodynamic regime earlier in Earth history [69, 70]. Extreme ultrahigh-pressure (UHP) metamorphic rock complexes are generated and exhumed by Phanerozoic collisional orogenic systems, which also create clockwise metamorphic P–T routes. About a thousand high-pressure (HP)-ultrahigh-pressure (UHP) metamorphic terrains have been discovered around the world, the most of which are Phanerozoic in age. One is Neoproterozoic, while the other is Neoarchean to Paleoproterozoic in age [1, 71]. The lack of UHP metamorphic complexes in the Precambrian geological record indicates that another type of orogenesis predominated earlier in Earth's history [64, 65]. Based on field results, Precambrian orogens differ greatly from current orogens. At high apparent geothermal gradients, Precambrian orogenesis made important contributions to crustal growth and magmatism [64, 66]. Four orogen categories was proposed by Chardon et al. [67] and Cagnard et al. [72] recently attempted to define Precambrian accretionary orogens using first-order structural and metamorphic traits, which represent the state of the continental lithosphere in these convergent settings involving enormous juvenile magmatism (**Figure 8**).

5.3 Micro-Raman spectroscopy

Figure 9 revealed Raman spectra were observed within the white matrix, pyroxene, opaque mineral pockets and diverse places around the mineral matrix. Maghemite Raman shift peaks are recorded at some points within the biotite granite gneiss, and thin section petrography of all the rock units in the study area shows the abundance of quartz, microcline and plagioclase as the major minerals that dominate the rock samples with other minor components such as hornblende, muscovite

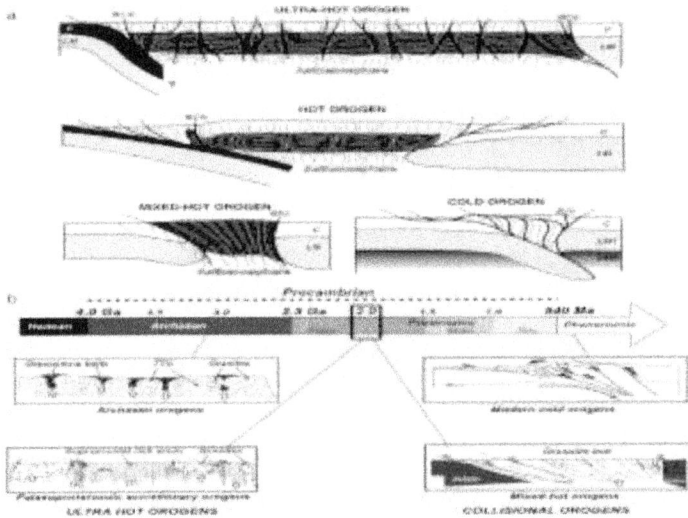

Figure 8.
Proposed classifications for Precambrian orogens Chardon et al. [67] (a) and Cagnard et al. [72] (b). (a) Orogen construction possibilities [67]: LM1 = stiff upper mantle lithosphere; LM2 = ductile, lower viscosity, lower lithospheric mantle; C = crust; LM = lithospheric mantle; LM1 = stiff upper mantle lithosphere; LM2 = ductile, lower viscosity, lower lithospheric mantle. (b) Schematic orogenic cross sections depicting the evolution of distinct orogenic styles over time. [72].

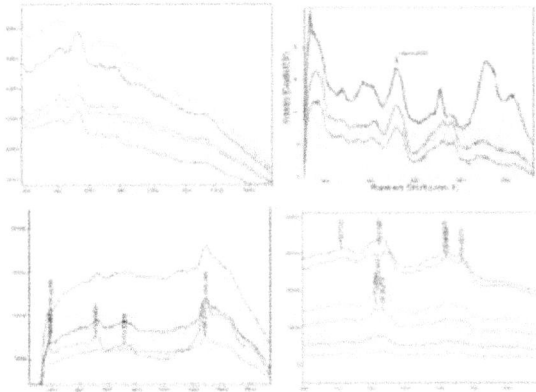

Figure 9.
Micro-Raman spectroscopy of some Precambrian rocks.

and the opaque minerals. Plagioclase, quartz, and microcline minerals were found to make up to 70% of the volume percentages of the rock in thin sections, with plagioclase being the most dominant mineral, followed by quartz microcline being the third most dominant mineral. The results of 830 nm Raman microspectroscopy of biotite granite gneiss grains have 398.8 cm⁻¹ and 663 cm⁻¹ indicating weak spectra while that of 785 nm have 714.8 cm⁻¹, 720.4 cm⁻¹ and 764.48 cm⁻¹ indicating strong peak bands respectively (**Figure 9**) [47, 73–75].

Raman shifts are: 398.8 cm⁻¹, 521 cm⁻¹, and 714.8 cm⁻¹ signals in P4332 mode, parallel swinging of the two tetragonal centres corresponds to polymorph Fe-O bond stretching, cubic P4332 structure, and suggested change in the tetragonal symmetry. This originates from Fe$_2$-, Fe$_3$- and O$_2$- which are the bond stretching in the [γ- Fe_2O_3] cubic P4332/tetragonal P41212 having transformation phase of magnetite spinel of α- Fe_2O_3 and γ- Fe_2O_3 polymorphs [76]. Kaiser microprobe of 785 nm recorded Raman peaks at 522.67 cm^{-1} and 714.8 cm^{-1}, 720.4 cm^{-1}, 764.48 cm^{-1} for maghemite mineral grains observed in biotite granite gneiss.

Low-temperature shortage causes oxidation of magnetite in the magnetic moment, resulting in a reverse spinel structure with both Fe_2_ and Fe_3_ ions in tetrahedral positions (A) and octahedral (B) sites configuration were all factors that contributed to the presence of maghemite in the biotite granite gneiss [76]. The main minerals in granite are quartz, microcline and plagioclase, while minor minerals include hornblende, muscovite, and opaque minerals. Under plane-polarised light, the quartz mineral in the rock samples was colourless, and it appears as subhedral prismatic crystals. Microcline is typified by polysynthetic twinning in two directions (cross-hatched), one according to albite law, and the other according to pericline law (monoclinic orthoclase/sanidine transformed to microcline), whereas its polysynthetic twinning distinguished plagioclase according to albite law. Biotite is brown, yellowish-brown and reddish-brown in the thin section. It is pleochroic, occurring as plates and laths and showed elongation along the foliation plane.

Several granite grains with variable colours were subjected to two excitation wavelengths of Raman microprobe, disregarding configuration and cause of the 521 cm^{-1} peak. The Raman mode at 519.1 cm^{-1} and 521 cm^{-1} corresponds to polymorph Fe-O bond stretching, cubic P4332 structure, which described the tetragonal distortion symmetry. The Raman shifts of 519 cm^{-1}, 522.67 cm^{-1}, 663 cm^{-1} and 714.8 cm^{-1} are laser excitation wavelengths and not fluorescence induced. These bands appeared only for dark or opaque granite crystals when excited with Raman microprobe (785 and 830 nm) laser, even if the clear samples are less intense. Since the points are similar to those determined for maghemite lattice modes, and these spectra correspond to translational lattice modes in maghemite geometry [73].

Scanning electron microprobe mineralogical compositions of granitoids from the basement complex of Southwestern Nigeria were studied using scanning electron microscopy (SEM). Gneiss, granite, biotite granite gneiss, banded gneiss and charnockite predominantly recorded maghemite/magnetite, ilmenite, pyrite and poor (titano)magnetites, with differences in titanium (Ti), grain sizes content and configuration, respectively (**Figure 9**). Examinations of polished sections of samples from the southwestern Nigerian granitoids (**Figure 10A**-F) revealed: grains of maghemite and magnetite (light grey); titanomagnetite (grey) magnetite (light grey); magnetite and titanomagnetite (striations of light grey); maghemite and titanomagnetite (grey); ilmenite and pyrite observed between (white and grey) respectively.

Scanning electron microscope serves as a proxy in determining the diverse magnetic phases in iron titanium oxides present in the selected rock samples. Studies showed larger altered (titano)magnetite grains in the gneiss, titanomagnetite in granite, phases of titanomagnetite and magnetite in biotite granite gneiss with evidence of dehydration, maghemite and titanomaghemite in banded gneiss, ilmenite in charnockite and pyrite as seen in granite gneiss (**Figure 10** a-d). The Fe-Ti-O grains indicated transformations of spinel rods, low-temperature oxidation reaction, precipitation of crystalline rock phases (exsolution) and dehydration due to tectonic-metamorphic episodes observed in studied rock samples. The abundance of precursor magnetite was susceptible to transformation than the smaller magnetite grains resulting in the pronounced formation of maghemite and titanomaghemite in the rock samples. This result correlates with the arguments of Carporzen *et al.* [77] that suggested tectono-metamorphism-related temperatures in the rock assemblage and heating of magnetite.

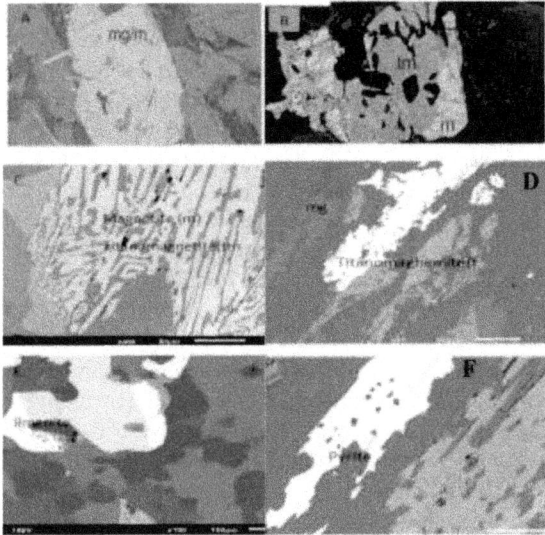

Figure 10.
SEM results of magnetic minerals (JEOLJSM-5900LV). (a) Altered magnetite (maghemite) of gneiss (scale:10 pm), (b) titano-magnetite of granite (scale:10 pm). (c) Titano-magnetite and magnetite from dehydration of biotite in biotite granite gneiss (d) titanomaghemite and maghemite of banded gneiss (e) ilmenite of charnockite (f) pyrite of Charnockite.

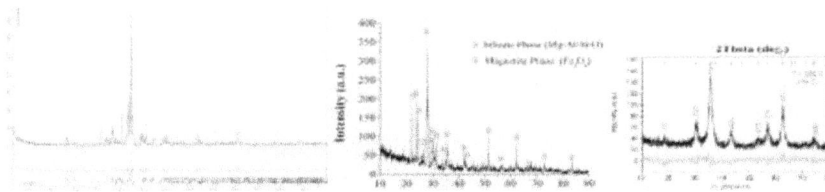

Figure 11.
XRD of some Precambrian basement rocks the study area (a) granite (b) gneiss (c) Charnockite.

5.4 X-ray diffractometry

The magnetic minerals in the selected rock samples were investigated further using the XRD data (**Figure 11** a-c) to unravel the mineralogical phases found in the granite, gneiss and charnockite. It demonstrates mainly the silicate and pseudomorphs of magnetite phases. **Figure 11** a-c showed pronounced spectra of silicate phase while the smaller spectra are pseudomorphs of magnetite (magnetite and maghemite). These results are consistent with that of the Raman spectroscopy, SEM and temperature dependence.

6. Discussion of results

The Precambrian rocks witnessed remagnetisation due to four phases of tectonometamorphic episodes. Rocks like diorite, biotite granite gneiss and syenite were not remagnetised and recorded normal and reversed polarities. The documentation of preliminary paleomagnetic, geochronological and microstructural datasets for Precambrian granitoids from the Southwestern Nigeria basement complex, located on the Pan African nappe system covered the Southwestern area of Nigeria, promoted extensive greenschist hydrothermal metamorphism in the underlying cratonic basement, according to research conducted on the northern edge of the Congo craton. The tectonometamorphism has resulted in widespread remagnetisation of the granitoids in southwestern Nigeria. On amphibole grains from Precambrian rocks, 206Pb/207Pb techniques were used to date both metamorphism and magnetic resets at 571.6 Ma. The normal and reverse polarities found in late Neoproterozoic granitoids are coeval with the paleopole at 304.80E and 61.80S (DP = 5.4, dm = 10.7) meet the fifth criteria of Van der vool. These pole and specific primary poles of the Congo craton propose an elbow-shaped apparent polar wandering path ranging from 593 to 547 Ma at the Pan African tectonic metamorphism [23, 26, 42]. Raman spectroscopy revealed the presence of maghemite iron oxide minerals in most of the rocks. SEM results showed maghemite, magnetite, titanomagnetite, ilmenite and pyrite, while XRD recorded pseudomorphs of magnetite. Two-sided lithospheric downwellings and shallow underthrusting weakened the plates. When the upper mantle temperature rises over 250°C, a "no-subduction" zone emerges, in which small deformable plate pieces move horizontally. The degree of lithospheric weakening caused by the intrusion of sub-lithospheric melts into the lithosphere controls the tectonic regime. At upper-mantle temperatures of 175–160°C, a reduced melt flow leads in less melt-related weakening and more strong plates, stabilising the present subduction type even at high mantle temperatures.

7. Conclusions

Remagnetization was prevalent in the Precambrian era. Reactivation of the Pan-African tectonics on the migmatite gneiss protolith (Eburnean granitic pluton) was not affected in numerous sites. The rocks demonstrated primary and secondary remagnetisation (normal, reversed and mixed polarities) and stability established by representative rocks of biotite granite gneiss, granite gneiss and syenite. Geochemical and isotopes parameters have revealed that the Paleoproterozoic/Eburnean orthogneiss and the granite plutons represent the same lithospheric source Paleoproterozoic/Eburnean source became molten all through the Pan-African event. The paleomagnetic pole positions of some Precambrian rocks in southwestern to the orogenic events revealed actual polar wander paths towards the equator during the assemblage of the Rodinia supercontinent. The Raman spectra of maghemite through estimation and observations of analogous wavenumbers of magnetite pseudomorphs revealed its atomic origin. The individual specimens of biotite granite gneiss, granite and charnockite have maghemite at strong peak spectra 519, 521, 522.00 cm^{-1} and 1285.5 and weak shoulder Raman spectra 398.8, 663, 710 and 717 cm^{-1} with 830 and 785 nm infra-red Raman spectroscopy. SEM revealed evidence of magnetite and titanomagnetite and dehydration of biotite in biotite granite gneiss. Study area dominated by plume tectonics and lithospheric delamination. Numerical models suggest that the transition occurred at mantle temperatures 175–250°C higher than present-day values triggered by stabilisation of rheologically strong plates of continental and oceanic type. Widespread development of modern-style (cold) collision on Earth started during Neoproterozoic at 600–800 Ma decoupled and is thus the onset of modern-style subduction. The cold collision created favourable conditions for the generation of ultrahigh-pressure (UHP)metamorphic complexes in southwestern Precambrian rocks.

Geodynamics of Precambrian Rocks of Southwestern Nigeria
DOI: http://dx.doi.org/10.5772/intechopen.104668

Author details

Cyril C. Okpoli[1,3*], Michael A. Oladunjoye[2] and Emilio Herrero-Bervera[3]

1 Faculty of Science, Department of Earth Sciences, Adekunle Ajasin University, Ondo State, Nigeria

2 Department of Geology, University of Ibadan, Ibadan, Nigeria

3 Paleomagnetics and Petrofabrics Laboratory, School of Ocean and Earth Science and Technology (SOEST), Hawaii Institute of Geophysics and Planetology (HIGP), Honolulu, Hawaii, USA

*Address all correspondence to: cyril.okpoli@aaua.edu.ng

IntechOpen

References

[1] Gerya T. Precambrian geodynamics: Concepts and models. Gondwana Research. 2012;**25**(2):442-463. DOI: 10.1016/j.gr.2012.11.008

[2] DePaolo DJ, Cerling TE, Hemming SR, Knoll AH, Richter FM, Royden LH, et al. Origin and evolution of earth: Research questions for a changing planet. Committee on grand research questions in the solid-earth sciences. In: Board on Earth Sciences and Resources, Division on Earth and Life Studies, National Research Council of the National Academies. Washington, D.C.: The National Academies Press; 2008. p. 137

[3] Benn K, Mareschal J-C, Condie KC. Archean geodynamics and environments. Geophysical Union. : Geophysical Monograph Series. 2006;**64**:320

[4] Van Kranendonk MJ. Onset of plate tectonics. Science. 2011;**333**:413-414

[5] Caby R. Terrane assembly and geodynamic evolution of central–western Hoggar: A synthesis. Journal of African Earth Sciences. 2003;**37**:133-159

[6] Ferre E, Gleizes G, Caby R. Obliquely convergent tectonics and graniteemplacement in the trans-Saharan belts of eastern Nigeria: A synthesis. Precambrian Research. 2002;**11**:199-219

[7] Caby R, Betrand JML, Black R. Pan-African ocean closure and continental collision in the Hogger-Iforas segment, central sahara. In: Kroner A, editor. Precambrian Plate Tectonics. Amsterdam: Elsevier; 1981. pp. 407-437

[8] Adetunji A, Olarewaju VO, Ocan OO, Macheva L, Ganev VY. Geochemistry and U-Pb zircon geochronology of Iwo quartz potassic syenite, southwestern Nigeria: Constraints on petrogenesis, timing of deformation and terrane amalgamation. Precambrian Research. 2018;**307**:125-136

[9] Dada SS. Proterozoic evolution of Nigeria. In: Oshin O, editor. The Basement Complex of Nigeria and its Mineral Resources (a Tribute to Prof. M. A. O. Rahaman). Ibadan: Akin Jinad & Co.; 2006. pp. 29-44

[10] Black R, Liégeois JP. Cratons, mobile belts, alkaline rocks and continental lithospheric mantle: The pan-African testimony. Journal of Geological Society London. 1993;**150**:89-98

[11] Liégeois JP, Latouche L, Boughrara M, Navez J, Guiraud M. The LATEA metacraton (central Hoggar, Tuareg shield, Algeria): Behaviour of an old passive margin during the pan-African orogeny. Journal of African Earth Science. 2003;**37**:161-190

[12] Ajibade AC, Wright JB. The Togo-Benin-Nigeria shield: Evidence of crustal aggregation in the pan African belt. Tectonophysics. 1989;**65**:125-129

[13] Caby R, Boesse JM. Pan-African nappe system in south-West Nigeria: 'Ife lle-llesha Schist Belt. Journal African. Earth Sciences. 2001;**33**(2):211-225

[14] Ferre E, Deleris J, Bouchez JL, Lar AU, Peucat JJ. The pan-African reactivation of contrasted eburnean and Archaean provinces in Nigeria: Structural and isotopic data. Journal of the Geological Society London. 1996;**153**:719-728

[15] Rahaman MA. Recent advances in the study of the basement complex of Nigeria. In: Oluyide PO, Mbonu WC, Ogezi AE, Egbuniwe IG, Ajibade AC, Umeji AC, editors. Precambrian Geology of Nigeria. Lagos, Nigeria: Geological Survey of Nigeria special publication; 1988. pp. 11-41

[16] Tubosun IA, Lancelot JR, Rahaman MA, Ocan O. U-Pb Pan African ages of two charnockite-granite association from southwestern Nigeria. Contribution to Mineralogy and Petrology. 1984;**88**:188-195

[17] Dada SS. Crust-forming ages and Proterozoic crustal evolution in Nigeria: a reappraisal of current interpretations. Precambrian Research. 1998;**87**:65-74

[18] Kröner A, Ekwueme BN, Pidgeon RT. The oldest rocks in West Africa: SHRIMP zircon age for early archean migmatitic orthogneiss at Kaduna, Northern Nigeria. Journal of Geology. 2001;**109**:399-406

[19] Okonkwo CT, Ganev VY. U-Pb zircon geochronology of the Jebba granitic gneiss and its implications for the Paleoproterozoic evolution of Jebba area, southwestern Nigeria. International Journal of Geoscience. 2012;**3**:1065-1073

[20] Salminen J, Klein R, Mertanen S. New rock magnetic results for the 1.64 Ga suomenniemi dyke swam, SE Finland. Precambrian Research. 2019;**329**:195-210

[21] Nzenti JP, Barbey P, Macaudiere J, Soba D. Origin and evolution of the late Precambrian high-grade Yaounde gneisses (Cameroon). Precambrian Research. 1988;**38**:91-109

[22] Rahaman MA. Review of the basement geology of southwestern Nigeria. In: Kogbe CA, editor. Geology of Nigeria. Lagos: Elizabethan Publication Company; 1976. pp. 41-48

[23] Sayab M, Lahtinen R, Koykka J, Holtta P, Karinen J, Niiranen T, et al. Improved resolution of Paleoproterozoic orogenesis: Multidirectional collision tectonic in the sodankyla belt of northern Finland. Precambrian Research. 2021;**359**:106193

[24] Black R, Caby R, Moussine-Pouchkine A, Bayer R, Betrand JM, Boullier MM, et al. Evidence for late Precambrian plate tectonics in West Africa. Nature. 1979;**278**(5701):223-227

[25] Obiora SC. Geology and Mineral Resources of the Precambrian Basement Complex of Nigeria. A talk presented at the National Geophysical Research Institute, Hyderabad, India on 25th November 2008 as a CSIR-TWAS Postdoctoral Fellow; 2008, p. 30

[26] Kalsbeek F, Affaton P, Ekwueme B, Freid R, Thranea K. Geochronology of granitoid and metasedimentary rocks from Togo and Benin, West Africa: Comparisons with NE Brazil. Precambrian Research. 2012;**196-197**:218-233

[27] Onyeagocha AC, Ekwueme BN. Temperaturepressure distribution patterns in metamorphosed rocks ofthe Nigerian basement complex a preliminary analysis. Journal African Earth Sciences. 1990;**11**:83-93

[28] Black RL, Latouche J, Liegeois C, R. and Bertrand, J. M. Pan-African displaced terranes in the Tuareg shield (Central Sahara). Geology. 1994;**22**(7):641-644

[29] Nigerian Geological Survey Agency (NGSA). Geological Map of Nigeria 1:250,000. Abuja Nigeria; 2010

[30] Pisarevsky SA, Wingate MTD, Powell CM, Johnson S, Evans DAD. Models of Rodinia assembly and fragmentation. Geological Society, London, Special Publications. 2003;**206**(1):35-55

[31] Gray R, Pysklywec R. Geodynamicmodels of Archean continental collision and the formation of mantle lithosphere keels. Geophysical Research Letters. 2010;**37**:L19301

[32] Eagles G. New angles on South Atlantic opening. Geophysical Journal International. 2007;**168**(1):353-361

[33] Meert J, Lieberman B. The Neoproterozoic assembly of Gondwana and its relationship to the Ediacaran-Cambrian radiation. Gondwana Research. 2008;**14**:5-21

[34] Dada SS. Proterozoic evolution of the Nigeria-Boborema province. Geological Society, London, Special Publications. 2008;**294**(1):121-136

[35] van Schmus WR, Oliveira EP, da Silva Filho AF, Toteu SF, Penaye J, Guimaraes IP. Proterozoic links between the Borborema Province, NE Brazil, and the central African Fold Belt. Geological Society, London, Special Publications. 2008;**294**(1):69-99

[36] Neves SP, Bruguier O, Bosch D, da Silva JMR, Mariano G. U-Pb ages of plutonic and metaplutonic rocks in southern Borborema Province (NE Brazil): Timing of Brasiliano deformation and magmatism. Journal of South American Earth Sciences. 2008;**25**(3):285-297

[37] Toteu SF, Penaye J, Djomani YP. Geodynamic evolution of the pan-African belt in Central Africa with special reference to Cameroon. Canadian Journal of Earth Sciences. 2004;**41**:73-85

[38] Abdelsalam MG, Liegeois JP, Stern RJ. The Saharan Metacraton. Journal of African Earth Sciences. 2002;**34**(3-4):119-136

[39] Attoh K, Hawkins D, Bowring S, Allen B. U-Pb zircon ages of gneisses from the pan-African Dahomeyide orogen, West Africa. EOS Transactions of the American Geophysical Union. 1991;**72**:299

[40] Hirdes W, Davis DW. U-Pb zircon and rutile metamorphic ages of Dahomeyan garnet-hornblende gneiss in southeastern Ghana, West Africa. Journal of African Earth Sciences. 2002;**35**(3):445-449

[41] Valeriano CM, Pimentel MM, Heilbron M, Almeida JCH, Trouw RAJ. Tectonic evolution of the Brasilia Belt, Central Brazil, and early assembly of Gondwana. Geological Society, London, Special Publications. 2008;**294**(1):197-210

[42] Chaves ADO. Columbia (Nuna) supercontinent with external subduction girdle and concentric accretionary collisional and intracontinental orogens permeated by large igneous provinces and rocks. Precambrian Research. 2021;**352**:106017

[43] Klein EL, Moura CAV. Sao Luis craton and Gurupi Belt (Brazil): Possible links with the west African craton and surrounding pan-African belts. Geological Society, London, Special Publications. 2008;**294**(1):137-151

[44] Chadima M, Hrouda F. Remasoft 3.0 a user friendly paleomagnetic data browser and analyser. Travaus Geophysiques. 2006;**27**:20-21

[45] Sagnotti L. Demagnetization analysis in excel (DAIE). An open source workbook in excel for viewing and analysis in demagnetization data from Plaeomagnetic discrete samples. Annals of Geophysics. 2013;**56. 1**:0114

[46] Fisher R. Dispersion on a sphere. Proceedings of Royal Society. 1953;**217**:295-305

[47] de Faria DLA, Silva SV, de Oliveira MT. Raman microspectroscopy of some iron oxides and oxyhydroxides. Journal of Raman Spectroscopy. 1997;**28**:873-878

[48] Shebanova ON, Lazor P. Raman study of magnetite (Fe_3O_4): Laser-induced thermal effects and oxidation. Journal of Raman Spectroscopy. 2003;**34**:845-852

[49] Chen LD, Heslop AP, Roberts L, Chang X, Zhao HV, McGregor G, et al.

Remanence acquisition efficiency in biogenic and detrital magnetite and recording of geomagnetic paleointensity. Geochemistry Geophysics Geosystem. 2017;**18**:1-17. DOI: 10.1002/2016GC006753

[50] Van Der Voo R. The reliability of paleomagnetic data. Tectonophysics. 1990;**184**(1):1-9

[51] Halla J, van Hunen J, Heilimoc E, Hölttäd P. Geochemical and numerical constraints on Neoarchean plate tectonics. Precambrian Research. 2009;**174**:155-162

[52] Moyen J-F, van Hunen J. Short-term episodicity of Archaean plate tectonics. Geology. 2012;**40**:451-454

[53] van Hunen J, van den Berg A. Plate tectonics on the early earth: Limitations imposed by strength and buoyancy of subducted lithosphere. Lithos. 2008;**103**:217-235

[54] van Thienen P, van den Berg AP, Vlaar NJ. Production and recycling of oceanic crust in the early earth. Tectonophysics. 2004;**386**: 41-65

[55] Albarede F, Van der Hilst RD. Zoned mantle convection. Philosophical Transactions of the Royal Society A. 2002;**360**(1800):2569-2592

[56] Van der Hilst RD, Karason H. Compositional heterogeneity in the bottom 1000 km of the Earth's mantle: Toward a hybrid convection model. Science. 1999;**283**:1885-1888

[57] O'Neill C, Lenardic A, Moresi L, Torsvik TH, Lee C-TA. Episodic Precambrian subduction. Earth and Planetary Science Letters. 2007;**262**:552-562

[58] Stein M, Hofmann AW. Mantle plumes and episodic crustal growth. Nature. 1994;**372**:63-68

[59] Condie K. Supercontinents and superplume events: Distinguishing signals in the geologic record. Physics of the Earth and Planetary Interiors. 2004;**6**:319-332

[60] Sizova E, Gerya T, Brown M, Perchuk LL. Subduction styles in the Precambrian: Insight from numerical experiments. Lithos. 2010;**116**: 209-229

[61] Davies GF. Gravitational depletion of the early Earth's upper mantle and the viability of early plate tectonics. Earth and Planetary Science Letters. 2006;**243**:376-382

[62] Davies GF. On the emergence of plate-tectonics. Geology. 1992;**20**:963-966

[63] Ueda K, Gerya T, Sobolev SV. Subduction initiation by thermal–chemical plumes. Physics of the Earth and Planetary Interiors. 2008;**171**:296-312

[64] Brown M. Metamorpjic conditions in orogenic belts: A record of secular change. International Geology Review. 2007;**49**(3):193-234

[65] Brown M. Characteristic thermal regimes of plate tectonics and their metamorphic imprint throughout earth history. In: Condie KC, Pease V, editors. When Did Earth First Adopt a Plate Tectonics Mode of Behavior? Vol. 440. Boulder, Colorado, USA: The Geological Society of America, Special paper; 2008. pp. 97-128

[66] Condie KC. Accretionary orogens in space and time. Geological Society of America Memoir. 2007;**200**: 145-158

[67] Chardon D, Gapais D, Cagnard F. Flow of ultra-hot orogens: A view from the Precambrian, clues for the Phanerozoic. Tectonophysics. 2009;**477**:105-118

[68] Windley BF. Proterozoic collisional and accretionary orogens. In: Condie KC, editor. Proterozoic Crustal Evolution. Amsterdam: Elsevier; 1992. pp. 419-446

[69] Gerya T. Future directions in subduction modeling. Journal of Geodynamics. 2011;**52**:344-378

[70] Van Hunen J, Allen MB. Continental collision and slab break-off: a comparison of 3-D numerical models with observations. Earth and Planetary Science Letters. 2011;**302**(1-2):27-37

[71] Liu S, Pan Y, Xie Q, Zhang J, Li Q. Archean geodynamics in the central zone, North China Craton: constraints from geochemistry of two contrasting series of granitoids in the Fuping and Wutai complexes. Precambrian Research. 2004;**130**(1-4):229-249

[72] Cagnard F, Barbey P, Gapais D. Transition between "Archaean-type" and "modern-type" tectonics: Insights from the Finnish Lapland Granulite Belt. Precambrian Research. 2011;**187**:127-142

[73] Chamritski I, Burns G. Infrared- and Raman-active phonons of magnetite, maghemite, and hematite: A computer simulation and spectroscopic study. Journal of Physical Chemistry B. 2005;**109**:4965-4968

[74] Hanesh M. Raman spectroscopy of iron oxides and (oxy)hydroxides at low laser power and possible applications in environmental magnetic studies. Geophysical Journal International. 2009;**177**:941-948. DOI: 10.1111/j.1365-246X.2009.04122.x

[75] Serna CJ, Rendon JL, Iglesias JE. Infrared surface modes in corundum-type microcrystalline oxides. Spectrochimica Acta Part A: Molecular Spectroscopy. 1982;**38**(7):797-802

[76] Dunlop DJ, Özdemir Ö. Rock Magnetism: Fundamentals and Frontiers. New York/London/Cambridge: Cambridge University Press; 1997. p. 573

[77] Carporzen L, Gilder SA, Hart RJ. Origin and implications of Verwey transitions in the basement rocks of the Vredefort meteorite crater, South Africa. Earth Planetary Science Letters. 2006;**251**:305-317

Exploring the Application of Potential Field Gravity Method in Characterizing Regional-trends of the Earth's Sequence System over the Sokoto Basin, NW, Nigeria

Adamu Abubakar and Othniel K. Likkason

Abstract

In this chapter some preliminaries evaluation were outline briefly based on exploring the use of potential gravity field method in characterizing regional trends of Earth's sequence system interms of gravitational potentials and fields, viz-a-viz., the background, instrumentation, theoretical model, gravity data reduction, geological framework, Bouguer reduction density, density acceptable models, free-air anomaly and interpretation of Bouguer gravity anomaly. Therefore, application of potential field method (gravity) explore these relationships by focusing on the formation and fill of a continental rift basin in characterizing regional trends of the Earth's system interms of data processing, interpretation and Earth's modelling. The study was carried out with the aim to understand and characterized the structural styles and regional trends of the Earth's sequence beneath the Sokoto Basin and its surroundings. Results from the Bouguer gravity anomaly revealed gravity high of characteristic feature on the Bouguer map with a strong positive oval shape of causative bodies anomaly (> -14.0 mGal) having E-W trend. On contrary a number of gravity minima (-43.31 mGal) and maxima (-39.54 mGal) can be found ENE, N-W parts which are almost defining locations of deep basinal areas. The anomalous features shows negative as well as the lineaments pattern are virtually oriented in the NW-SE, ENE and E-W trends.

Keywords: potential field, gravity method, regional trends, Earth's characterization, Sokoto Basin, interpretation model

1. Introduction

Earth scientists explore and investigate the structures of the Earth using diverse means, such as tectonic mapping, solid minerals, groundwater and hydrocarbon or for the harvest of geologic structures. Earth scientists may be interested in the determination of, for example, the thickness of sedimentary sequence, depth to basement structures and delineation of fractures (shallow and deep plate sources) for appropriate use in resource evaluation. For example, the identification and mapping of geometry, scale and nature of basement structures are critical in understanding the influence of basement during rift development, basin evolution and subsequent basin inversion. From regional gravity data, information such as tectonic frame work and other aforementioned information can be obtained. The geophysical information invariably combined with geological data are essential for a better understanding of the subsurface and characterizing regional trends of the Earth's structures. The use of gravity, can powerfully lead to a better detection and geological interpretation of structural features and has the potential of constraining quantitative details and reducing the ambiguity of geological interpretation. Geophysical method involving gravity are commonly used in the structural interpretation of sedimentary basins because of their better spatial resolution [1]. Potential field gravity method has proved very effective for providing useful information known to guide various exploration campaigns, be it regional studies, economic mineral or oil and gas exploration [2, 3].

Meaningful reconnaissance and detailed geological information have been generated by the analyses of gravity data for defining basin's tectonic framework, gravity survey is the primary method in geophysical exploration as a regional and local structural mapping tool [4–12]. The effectiveness of gravity survey depends on the existence of a significant density contrast between altered rocks or structures and their host rocks. Moreover, gravity survey not only reflects the shape of major granitoids, but also a correspondence between the tectonic lineaments and regional fault systems [12]. The present chapter guide and explore on the use of the acquired gravity data in characterizing regional trends of the Earth system in some parts of the sedimentary terrain of Africa (i.e. the Sokoto Basin of Nigeria, The Agnes of Egypt as well as Kenya). It's evidence that the gravity method depends on the different earth materials which have different bulk densities (mass) that bring out variations in the measured gravitational field. The variations can be interpreted through the use of enhancement techniques to determine the density, geometry and depth which causes the gravity variations in gravitational field. The Earth's gravitational field anomalies results from lateral variations of subsurface materials density and the distance from the measuring instruments, the general problem in geophysical surveying is the ambiguity in data interpretation of the subsurface geology. This arises because many different geologic configurations could reproduce similar

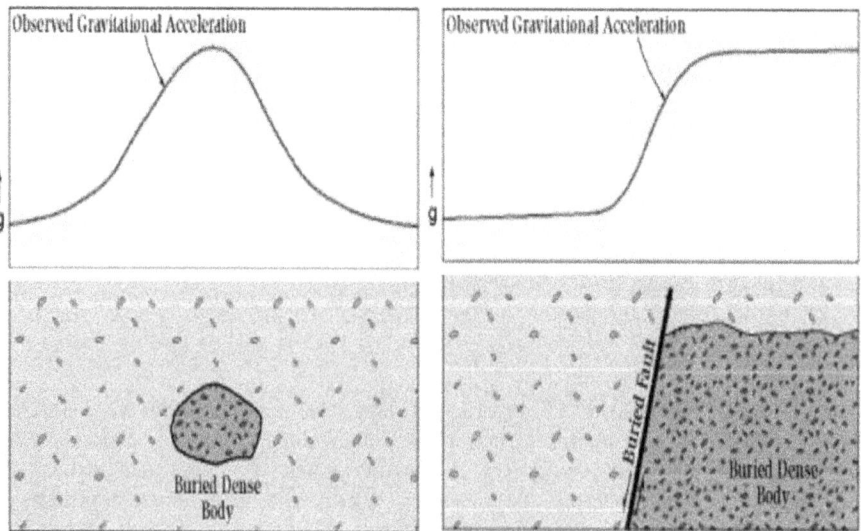

Figure 1.
Illustrations showing the relative surface variation of Earth's gravitational acceleration over geologic structures, after [13].

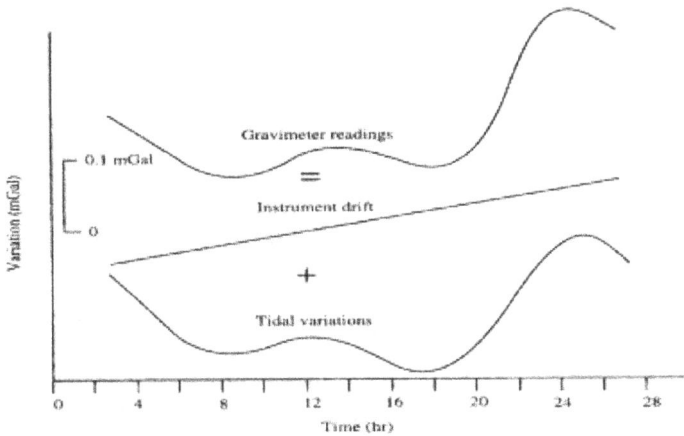

Figure 2.
Typical linear drift curve (middle curve) which is a combination of instrument drift and earth tidal variations.

observed measurements (**Figure 1**). The method can infer location of faults, permeable areas for tectonic movement. It is however, more commonly used in determining the location and geometry of Earth's system characterisation (**Figures 1** and **2**).

2. Instrumentation

The Lacoste and Romberg model gravity meter was used in data acquisition for this study. It has an advantage of repeatability of 3 mGal (980,000,000 mGal is the Earth's gravitational field) reading and is one of the preferred instruments for conducting gravity surveys in industry. It has a reading precision of 0.01 mGal and a drift rate less than 1 mGal per month (model G569 manual). Measurements were also made along designated areas to further check the behaviour of the instrument.

Figure 3.
The La Coste Romberg gravity meter.

All necessary routine checks on level adjustments and sensitivity of gravimeter were carried out as described in gravimeter manual (**Figure 3**).

3. Gravity data reduction

It's understood, the Earth's is slightly irregular oblate ellipsoid which means that the gravity field at its surface is stronger at the poles than the equator. The density distribution is irregular, particularly in an inelastic crust, which causes gravity to vary from expected value as the measurement position changes. Therefore, the variations are expressed as gravity anomalies. Mapping the gravity anomalies gives an understanding the structure of the Earth's [1, 5]. It's therefore essential to identify the reasons gravity varies and that it can be corrected while using gravity method in exploring and characterizing regional trends of the subsurface [5]. In this present survey, correction for the tide was not made because loops were closed at interval of about 2 hours or less. Also, since the area is relatively flat, there was no need considering excess mass or mass deficiency, hence terrain correction was not carried out. The results of gravimeter measurements are gravity differences between an arbitrary reference point and a series of field stations. The measured values at each station have some influences which completely mask the desired effect if they were not removed. Therefore before gravity measurements may be useful in possible indications of subsurface conditions (The observed gravity differences must be corrected for those various large influences). The objective of data is to remove the known effects caused by predictable features that are not of the target. The remaining anomaly is then interpreted in terms of subsurface variations in density. Each known effect is removed from observed data. The various corrections are described below:

3.1 The latitude correction

Both the rotation of the earth and its slight equatorial bulge produce an increase in gravity with increase in latitude (**Figure 4**). Therefore it becomes necessary to apply latitude correction for stations at different latitudes. The value of gravity increases with the geographical latitude [5]. With advance of Earth's rotation, the Earth's is not spherical but is flattened at poles thus the distance factor causes the 'g' value to increases from equator to pole by 6.6 Gals because the surface is closer to the centre at the poles (**Figure 4**) [1, 14]. The formula for latitude effect is the 1967 Gravity Reference System (GRS67) whose approximation is of the form:

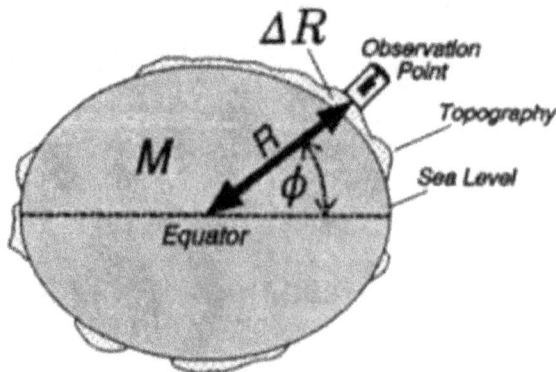

Figure 4.
Earth's rotation bulging at the equator.

$$g\theta = 978,171.261\ (1 + 0.005278895\sin^2\theta + 0.000023462\sin^4\theta)\ \text{mGal} \qquad (1)$$

Where θ is the latitude of the station concerned in degrees.

3.2 The free air correction

Free air anomaly is obtained from the difference between the measured or absolute gravity of a station, g_{obs} at the topography surface and its theoretical gravity, g_{lat}, extrapolated from the reference ellipsoid and correcting it for the free air effect. The final result of the free air anomaly is given as:

$$\Delta_{gyA} = g_{obi} - \left(g_{lat} - \frac{dg}{ds} h \right) \qquad (2)$$

where $\frac{dg}{ds}$ is the vertical gradient and it is the station elevation above mean sea level in meters and its value is 0.3086 Gal m^{-1} [1, 5]. In practice the value of 0.3086 mGal/m is the only value used after deriving from Eq. (2) thus, assumed that the Earth's is spherical and non-rotating. Finally, the correction considers only elevation differences relative to a datum and does not take into account that the mass between the observation point and datum as the station were suspended in free-air, not sitting on land (**Figure 5**). These serve as the reason that the correction termed as free-air correction (**Figure 5**). In general the datum used for gravity surveys is sea level and gravity decreases 0.3086 mGal for every meter above sea level [1, 5].

3.3 Bouguer correction

This is the difference between the observed gravity and the theoretical gravity at any point on the earth corrected for the mass of materials between the point and the datum plane (**Figure 6**), its value 0.04188 ρ, where ρ is the density of the slab [4, 5, 8, 11, 15]. Bouguer correction is applied in the opposite sense of free air that is it is subtracted when the station is above the datum plane and vice-versa. Bouguer correction accounts for gravitational of the mass above sea-level datum (**Figure 6**).

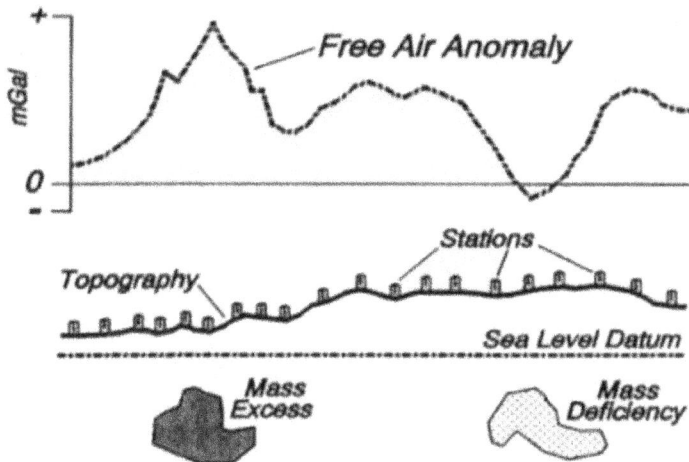

Figure 5.
Free air correction.

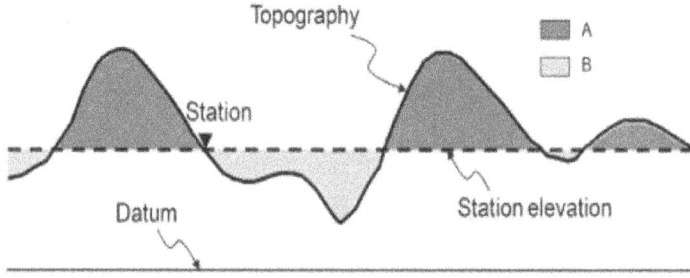

Figure 6.
Bouguer correction.

The equation for Bouguer gravity at a point after all the necessary preceding corrections have been applied can be written as:

$$g\left(B_{grav}\right) = g(obs) - BC \tag{3}$$

Where BC is the Bouguer correction.

$$g(BA) = g\left(B_{grav}\right) - g \text{ (Theoretical)} \tag{4}$$

Hence the Bouguer anomaly is determined using the expression:

$$\Delta G_{BA} = g_{obs} - g_{lat} + \frac{dg}{ds}h - 2\pi\rho Gh \tag{5}$$

Where $\pi\rho$ is the assumed crustal density value which is 2.67×10^{11} kgm^3 or Bouguer density and G is the universal gravitational constant. The term 2 Gh in the Bouguer correction which is the additional attraction exerted on a unit mass by a slab of rock material of density between a station and reference datum-mean sea level (m.s.1).

3.4 Bouguer gravity anomaly

A Bouguer gravity anomaly is the difference between the observed acceleration of an object in free fall (gravity) on surface of the Earth's, and the corresponding value predicted from a model of the gravitational field. If the attraction due to the effect of material between the plane of observation and the m.s.l. (known as the Bouguer correction (Bc)) is subtracted from the free-air anomaly, the corrected gravity field is called the Bouguer gravity anomaly and is given by:

$$g_B = g_{obs} + d_{gL} + d_{gFA} - d_{gB} - d_{gT} \tag{6}$$

Where g_{obs} = station readings; d_{gL} = latitude correction; d_{gFA} = free air correction; d_{gB} = Bouguer correction; d_{gT} = terrain correction. Putting in numerical values we have:

$$g_B = g_{obs} - g_0 + 0.3086H - 0.0419 \, rH - dgT \tag{7}$$

3.5 Drift correction

In the reduction of gravity data, the removal of drift which occurs as a result of elastic creep in the spring of the instrument is very necessary. The instrumental drift of the gravimeter used in this survey was removed using a Geosoft computer Algorithm routine of [13]. It is assumed that there is a linear relationship of the drift with time as given by the drift rate which is expressed as:

$$\mu = \frac{(g_2 - g_1) - (R_2 - R_1)}{t_2 - t_1} \tag{8}$$

Where g_1 and g_2 are absolute gravity values at the two end stations of a loop while R_1 and R_2 are tie observed reading (converted to milliGal) at times t_1 and t_2, respectively at those stations.

If the sums station is reoccupied, then $g_2 - g_1 = 0$ therefore Eq. (10) becomes; repeated computation for loops continued until all observations are referred to an initial time the drift correction for any intermediate station referred to the initial time t_O thus becomes:

$$\mu = \frac{(R_2 - R_1)}{t_2 - t_1} \tag{9}$$

With the assumption that drift of the instrument is a linear function of time over a short time interval, it was ensured that all observations in a day were tied to the same time origin during a day's work and the repeat observations at the same station after drift correction was equal to the former. Drift correction was done separately for each altimeter height value using free same cascade drift model. The absolute elevation for each of the stations were determined for each altimeter using the height of the Bench Mark No BM15 to which they were tied. Due to the characteristic behaviour of instrument [15], the field values recorded front tie altimeters for each station were varying. The observed gravity value at the detailed station is given by:

$$G_{obs} = g_1 - K\left[(R_o - R_1) - \mu(t_o - t_o)\right] \text{ mGals} \tag{10}$$

Where g_1 is the absolute gravity value at fee first base station, K is the meter constant. R_o and R_s, t_1 and t_o are the readings and times at the first base station and detail nation, respectively.

4. Geological framework

The Sokoto Basin is the Nigerian sector of the larger Iullemmeden Basin which spans parts of Algeria, Benin Republic, Niger Republic, Mali and Libya [16]. The study area falls within the Sokoto basin and lies between Latitudes 3:30 E–5:30 E and longitudes 11: 00 N–13:00 N. It is geographically located in the semi-arid with a zone of savannah-type vegetation as part of the sub-Saharan Sudan belt of West Africa with an elevation ranging from 250 to 400 m above sea level (**Figure 7a** and **b**). The area enjoys a tropical continental type of climate. Rainfall is concentrated in a short-wet season, which extends from April to October [17]. Mean annual rainfall is about 800–1000 mm while the mean annual temperature ranges from 26.5 to 40°C. Night temperatures are generally lower. The highest temperature occurs between April and July, the lowest in August (during the rainy season). An average nature of 40% low humidity during the wet season reaches a maximum of 80%, explain the dry nature of the environment in the area of study (**Figure 8**), which is in agreement of a sharp contrast to a humid environment when compare in the southern parts of Nigeria. The Sokoto Basin is predominantly a gentle undulating plain with an average

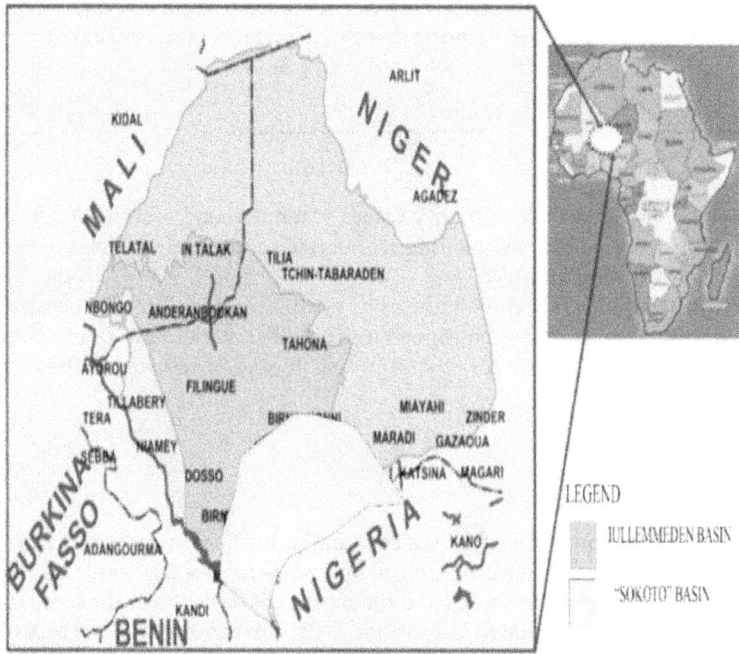

(a)

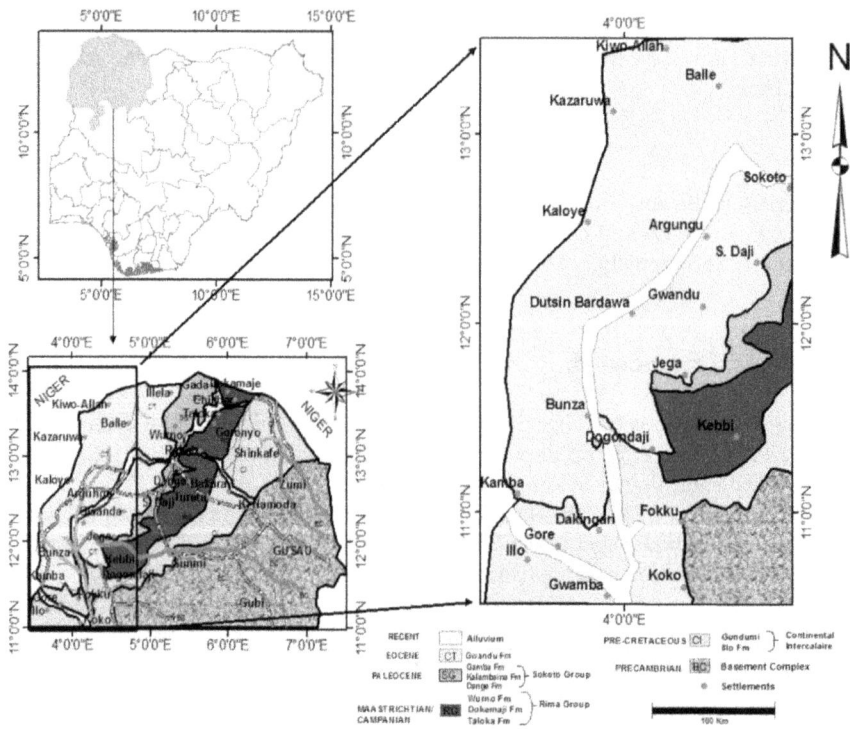

(b)

Figure 7.
(a, b) Geological map of Nigeria showing the "Sokoto Basin" and the study area [17].

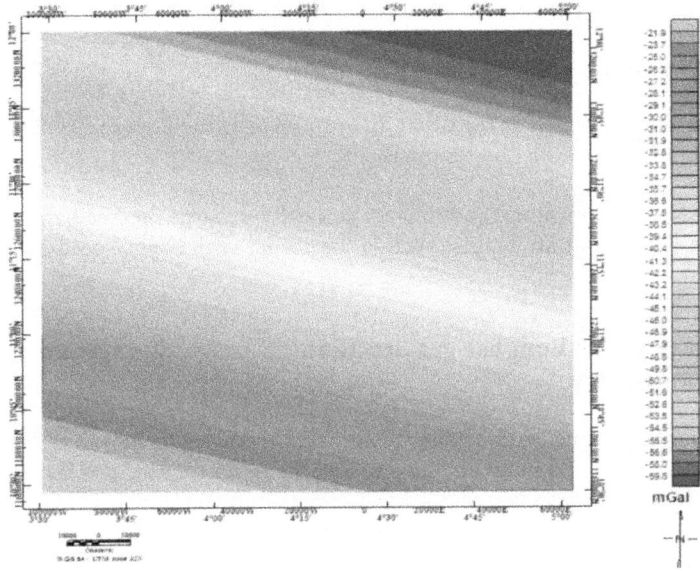

Figure 8.
Regional gravity anomaly map of the study area.

elevation varying from 250 to 400 m above sea level. The plain is occasionally interrupted by low mesas and other escarpment features [18–20]. The sediments of the Iullemmeden Basin were thought to accumulate during four main phases of deposition as follows (**Figure 7b** and **Table 1**):

a. The Illo and Gundumi Formations (made of grits and clays) unconformably overlie the pre-Cambrian Basement Complex. This is the so-called pre-Maastrichtian 'continental interclaire' of West Africa [16].

b. Next on the succession is the Maastrichtian (66–72 Ma) Rima Group (consisting of mudstones and friable sandstones) (the Taloka and Wurno formations) separated by the fossiliferous, calcareous and shaley Dukamaje Formation overlies the Illo and Gundumi Formations unconformably [17–19].

Age	Type/group	Formation	Sediment type	Remark
Eocene	—	Gwandu	Continental	Continental terminal
Paleocene	Sokoto Group	Gamba	Marine	—
		Kalambaina	Marine	—
		Dange	Marine	—
Maastrichtian	Rima Group	Wurno	Continental	—
		Dukamaje	Marine	—
		Taloka	Continental	—
Pre-Masstrichtian		Gundumi-Illo	Continental	Continental intercalaire
Pre-Cambrian	Basement complex rocks			

Table 1.
Stratigraphic successions in the Sokoto Basin, after [20].

    c. The Dange and Gamba Formations (mainly shales separated by the calcareous Kalambaina Formation constituting the Paleocene) (56–66 Ma) Sokoto Group overlie the Rima Group [18].

    d. The sequence cover is the Gwandu Formation of the Eocene (33–56 Ma) age forming the continental terminal [19].

The sediments dip gently and thicken gradually towards the northwest with maximum thicknesses attainable towards the border with Niger Republic.

5. Accuracy of the Bouguer gravity anomaly in Earth's system characterization

The computed Bouguer anomalies could have several errors introduced to it. Errors could be as a result of incompleteness of the formulae used and the correctness of the numerical values of the constants occurring in them [13, 15]. The calibration factor of the modern Lacoste and Romberg gravimeter depends only on the quality of the measuring screws and the lever system. Errors which could arise from the calibration factor is thought to be negligible because, the calibration factor does not change perceptibly with time, which eliminates the need for frequent checks of calibration. At each station, errors could arise from several sources. These include: errors in elevation determination (eh), errors in terrain effect (et), errors in base value (eb) errors is assumed which recommended that the most likely to in situ densities of subsurface rock lies between the dry and the saturated densities. The summary of the results for the various rock types identified in the area are shown in **Table 2**.

5.1 Bouguer reduction density

The objective of gravity survey is to detect subsurface density variations. Observed/measured gravity value at the station includes all kinds of attraction. Remove the effect of attraction except that of subsurface density anomaly (**Figure 9a**). There are three methods of selection of Bouguer reduction density; one is a "traditional" or standard density with which most regional maps have traditionally been reduced using a value of 2.67×10^3 kgm^{-3} (**Figure 9b**). The second is by determining a Bouguer reduction density which minimizes the correlation between the computed Bouguer anomaly and topography. This method is widely used in areas of rugged topography [21] and which was originally suggested by [11]

Rock type	No. of samples	Range of densities × $10^3 \times$ kgm^{-3}	Mean densities × $10^3 \times$ kgm^{-3}	Standard deviation
Limestone	63	1.87–2.99	2.43	0.16
Clay	45	1.46–3.62	2.54	0.09
Shales	22	1.76–5.30	3.53	0.04
Ironstone (finegrained)	43	1.97–3.50	2.74	0.06
Ironstone (coarse)	55	1.67–2.01	1.84	0.12
Total number of rock samples	228			

Table 2.
Summary of rock densities, after [14].

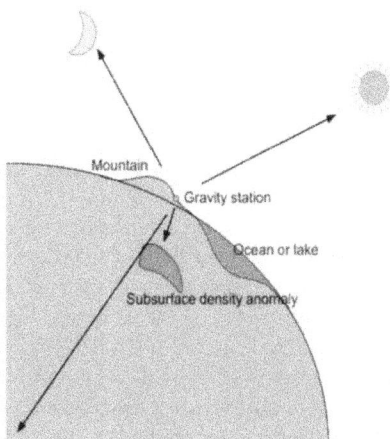

(a)

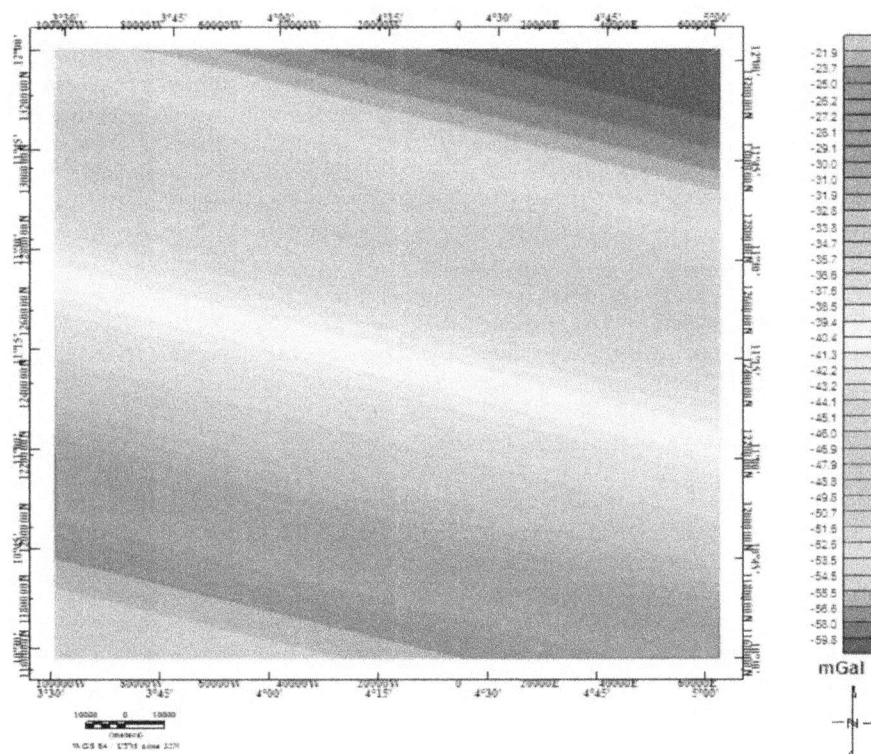

(b)

Figure 9.
(a, b): (a) Bouguer correction for subsurface density variations. (b) Regional gravity map at upward continue of 1 km after [14].

and [22]. This second method was not used in this chapter because the area is relatively flat. The third method is to measure the density of representative rock samples just as described and characterized them interms of earth system evolution. The fact is that it is usually difficult to obtain a suite of rock samples that is truly representative [2, 23]. Therefore in order to ensure consistency and compatibility with after regional gravity map in adjacent areas, the standard density value of 2.67×10^3 kgm^{-3} was used for reduction in this survey purpose.

5.2 Densities acceptable for models

The proper density values used for gravity interpretation depend upon the depth of formation in relation to the water table, which will in turn depend on whether the climate is arid or moist. The age and depths of sediments depend on how long they are buried. If the period is long enough, the sediments usually consolidate and lithify, resulting in a reduction in porosity and increase in density. Limestones and sandstones which are found in the study area increase in density by infiltration of cements without volumetric change [17, 18]. Clays and shales which are compacted clays are the most highly compressible of all sedimentary rocks and they therefore show the greatest amount of compaction. If sandstone and limestones on the other hand are subjected under the same environment, they experience smaller density change [18–20].

From the **Table 2**, it can be seen that the range for the rock density in some part of the Sokoto basin (i.e. Argungu, Shanga Kalambaina, Dange), area is from about 1.5×10^3 kgm^{-3} to 4.5×10^3 kgm^{-3} and their respective mean densities approximately agree with the published values for similar rock types from other places [1, 8]. Considering the results from Telford et al. for example, the mean densities for limestone, clay, shales, and laterites are (3.45, 2.43, 2.50 and 2.66) 10^3 kgm^{-3}, respectively and from the table, the same set of rocks have their density values ranging from 1.76 to 5.30×10^3 kgm^{-3}. Since limestone and laterites are the dominant rocks in the study area, density values within this range were used w that of the sediments during interpretation. Generally, the common rocks of the basement are gneisses, granites, phyllites and quartzite, their densities range from 1.67 to 2.01×10^3 kgm^{-3} and their average densities are (2.80, 2.64, 2.74 and 2.77) $\times 10^3$ kgm^{-3} as mentioned above, respectively [8]. Therefore the average density of the earth crust (2.67×10^3 kgm^{-3}) was then used as that of the basement.

5.3 Free-air anomaly map and topography

Free-air correction essentially takes care of the vertical decrease of the gravity with increase of Elevation and no account of the materials between the station and the datum plane taken. The variation amounts to -0.3086 mGal/m. The relationship between the free-air anomaly and heights was investigated and explained in the previous Section 5.1 above. The result of the free air anomaly are shown in **Figure 10** below. The free-air anomaly map indicates values ranging from a maximum of 11.8 mGal to a minimum of -41.2 mGal and a contour interval of 2 mGals was used for the map. A careful study of the map reveals that major linear pattern is generally in NE-SW direction with exception of few anomalies located at the S-W trend of the area.

5.4 Interpretation of Bouguer anomaly map

The Bouguer gravity map (**Figure 11**) comprises various low and high anomalies extend in the NW-SE, ENE and E-W trends as consequence with fold patterns in the southeastern part of Iullemmeden basin (Sokoto Basin in particular). These alternated anomalies are primarily due to the density contrast between the sedimentary blanket and some portion of the crystalline basement in Taloka formation. Sokoto Basin gravity high is a very characteristic feature on the Bouguer map with a strong positive oval shape anomaly (> -14.0 mGal) having S-E trend. The

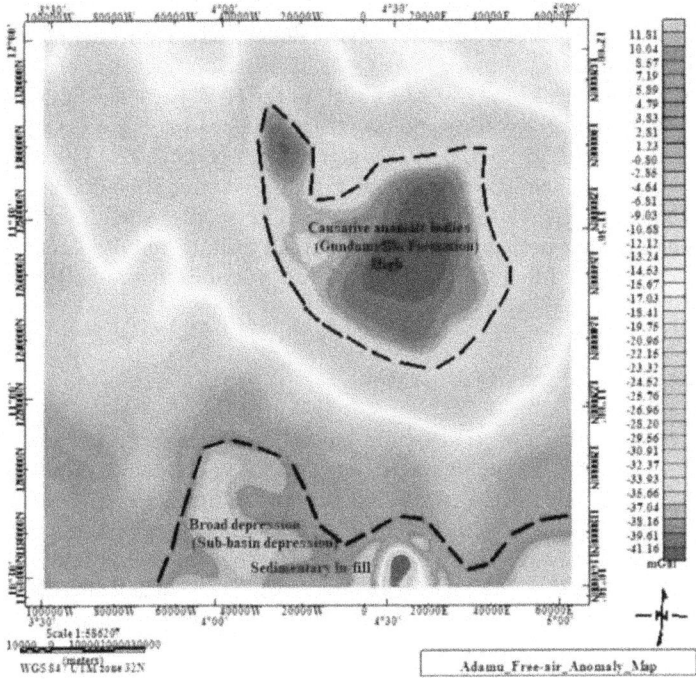

Figure 10.
Free air gravity anomaly map over Sokoto Basin after [14].

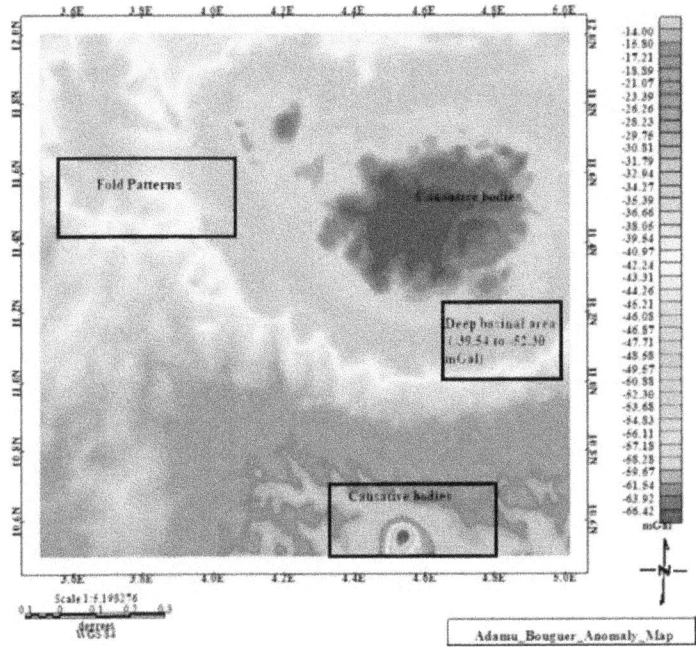

Figure 11.
Bouguer gravity anomaly map over Sokoto Basin after [14].

structural trend pattern from the map trending in the S-E direction is associated with deep basinal area of the causative (anomalous) body determined from the gravity survey are found to range from (−43.31 mGal) in the west corner of (**Figure 11**) fold patterns, and (−39.54 mGal) in the N-W part. The Bouguer gravity anomalous shows negative values and the structural lineaments patterns trending E-W (major trends) and NW-SE, ENW (minor trends) (**Figure 12**). The deep basinal of the causative (anomalous) body are fall within Gwandu, Kalambaina, Dange, Gamba, Wurno and Taloka formation represents two complementary different events; an older event probably of Continental intercalaire and pre-Cretaceous ages which caused major folding and faulting of NE-SW and ENE trends (**Figure 12**), respectively [17, 18].

5.5 Regional-residual gravity separation

In the present study, a purely analytical method was used with (Geosoft Oasis montaj V.8.4.3) in which matching of the regional by a polynomial surface of low order exposes the residual features as random errors. A first order polynomial surface was considered adequate for estimating the regional effect. Regional-residual separation process was applied to gravity data-set in order to estimate the amplitude of the regional background. Upward continuation was used to separate a regional gravity anomaly resulting from deep sources from the observed gravity (Bouguer anomaly) (**Figure 11**). The regional field (**Figure 9b**) is a plane dipping gently in a NW-SE direction with a gradient of about 1 mGal/km. The regional effect correspond to low frequencies therefore the anomalies are usually of long wavelength showing a gradual change in value while the residual anomalies which are due to local effects may show larger variations [6, 7]. There are several methods of removing the unwanted regional, some approach is entirely graphical while others are analytical. In some cases the graphical methods are incorporated in the analytical methods. The regional gravity values shows the negative entirely and are found to range from a maxima of −21.9 mGal to a minima of −59.3 mGal (**Figure 9b**).

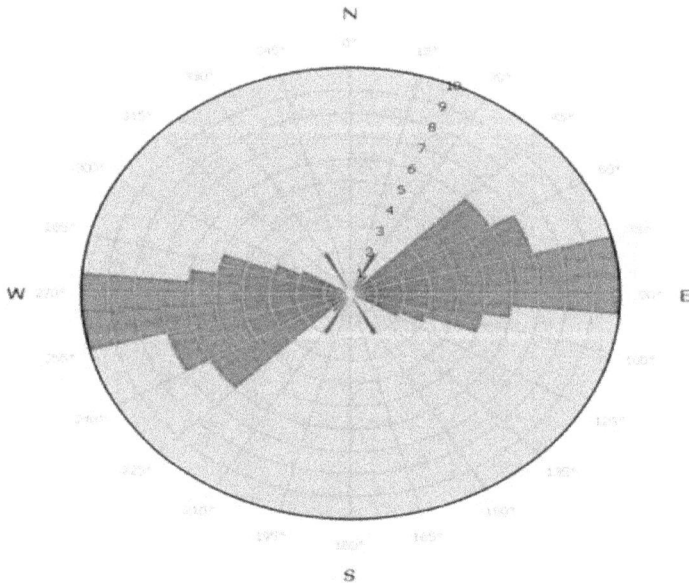

Figure 12.
Rose diagram, structural trends pattern.

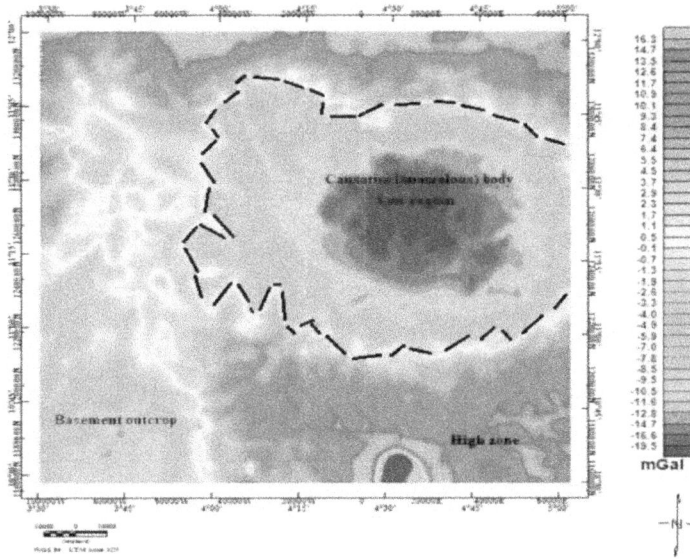

Figure 13.
Residual gravity map at upward continue of 1 km.

The residual anomaly at any point is then calculated as the difference between the observed Bouguer anomaly g_B and the regional effect g at that point and this is expressed as:

$$g_{res} = g_{OB} - g \qquad (11)$$

The residual anomalies at all the points were gridded and contoured through the application of (Geosoft Oasis montaj V.8.4.3). The resulting map (**Figure 13**) shows the gravitational effect of the near surface and local structures in the study area and the values was found to range from −19.5 mGal (minima) and 16.3 mGal (maxima). The larger features generally show up as trends which continue smoothly over very considerable areas, and they are caused by the deeper heterogeneity of the earth's crust superimposed on these trends, but frequently camouflaged by them, lie the smaller, local disturbances, which are secondary in size but primary in importance. These are the residual anomalies, which may provide the direct evidence for reservoir—type structure or mineral bodies.

6. Gravity data modelling/advanced processing

This process is aimed in modelling the source of the gravity signal measured at the surface. This can be done through processing of

- 2D density models

- 3D density models

As the Bouguer gravity value represents the effect of crustal and upper mantle density variations, the gravity anomalies were used to study the entire lithosphere. The 2D modelling (**Figure 14**). In quantitative interpretation of gravity data, the

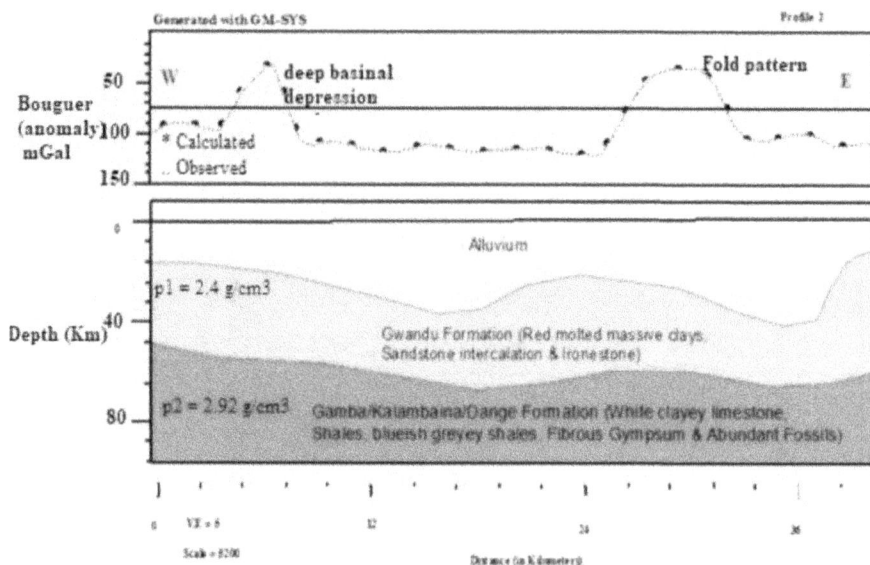

Figure 14.
2D regional crustal modelling of the Sokoto Basin, after [14].

objective is to estimate a subsurface structure whose calculated gravity effect satisfactorily approximate the observed gravity field measured on the surface. The magnitude of gravity anomaly caused by any structure depends directly on its volume times its density contrast. Secondly, the amplitude of the anomaly decreases as the depth of the structure causing it increases. If the shape of the structure is irregular or diffused, the observed gravity will be predictable to reduce in sharpness and in magnitude. Quantitative interpretation, generally barely unique or specific as it is always based on geologic implications. Thus, sufficient and adequate information about the geology of the study area becomes necessary for a meaningful interpretation. The study area falls into the sedimentary basin in the northwestern part of Nigeria, and the particular sediments found from the surface (Gwandu formation) which is of the Eocene age have average density of about 2.74×10^3 kgm^3 considering the lithologic sequence downward to depth of about 45 m. Underlying it with a slight unconformity are the Sokoto groups (Gamba, Kalambaina and Dange formation) which are of Paleocene age which have average density of about 2.43×10^3 kgm^3. These deposits extend to the depth of about 80 m (**Figure 14**) [18]. Below this occurs continental deposits (fluvial) which were of lower cretaceous or pre-Maastrichtian age. The estimated density value has a first density contrast of 2.4×10^3 kgm^3, and the second of 2.92×10^3 kgm^3, with respect to the average density of the basement (2.48×10^3 kgm^3) used. Therefore almost all the gravity lows in the study area were accounted for by the thickening of the sediments. In the interpretational procedures, the gravitational effect of any assumed initial model is calculated and compared with the observed effect. Changes are made as necessary on the presumed model in order to get a better fit. The common changes usually involve volume, shape and density contrasts. This process is repeated within geologically realistic limits until a new structure whose calculated effect best fits the observed effect was obtained. This approach is referred to as forward modelling (**Figure 14**). Profile 2 was chosen atleast to cross one major causative (anomalous) bodies identified earlier for interpretation (**Figure 13**). The computer program used for quantitative gravity interpretation of this profile 2. This

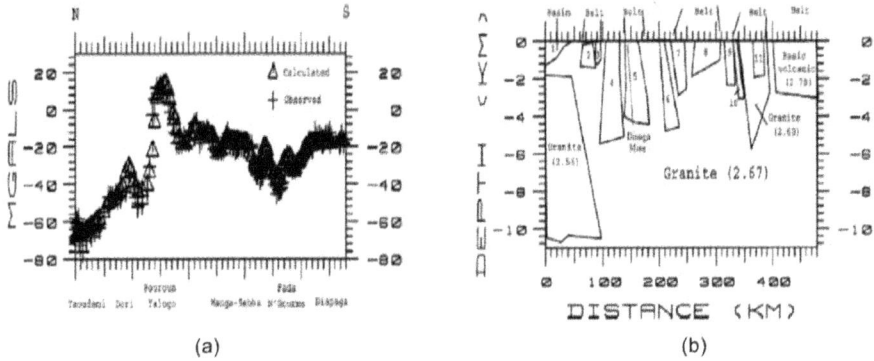

Figure 15.
(a, b) 2D density interpretation model.

interpretation reveals the prominent Gwandu formation and the Sokoto groups (i.e. Gamba, Kalambaina and Dange formation) of sedimentary in-fills have a common origin. The profile runs in the E-W direction and cuts across the causative (anomalous bodies) while modelling, an intrusion lie with density contrast of 2.43×10^3 kgm^3 introduced in part of Gwandu formation (**Figure 14**) at about 20–45 km along the profile before a fit of the computed with the observed was obtained in uppermost part of (**Figure 14**). While the low gravity at the western side of the profile was accounted for by thickened sediments which has high density contrast of 2.92×10^3 kgm^3. The maximum and minimum depths to the top of the sediments in-fills along this profile are 40 and 80 km, respectively. The body it has inward dipping walls and the dips are 45° and 55° on its western and eastern flanks, respectively has been calculated from the GYS-System.

6.1 2D density interpretation

In indirect interpretation, the Earth's sequence model whose theoretical anomaly can be computed simulates the causative body of a gravity anomaly characterization. The shape of the body can be altered until the computer anomaly closely matches the observed anomaly (**Figure 15a and b**).

6.2 Gravity contribution to conceptual model

Gravity methods are good in structural mapping in potential exploration for: Earth's system characterization interms of (imaging the lithospheric structures, dense material in shallow crust), fractures/faults (gravity gradients/slopes), help to identify potential drilling sites, help to identify potential recharge areas, etc.

7. Conclusion

The chapter were able to explore the potential application of gravity method for Earth's system exploration interms of regional trend characterization in African tectonic evolution settings. The strength of the potential gravity field method lies in the adequate density mass distribution of gravitization effect within the crustal materials of the Earth in the light of measurable gravity field over them. The Earth's gravitational field, that is the Earth's shape and global force, is itself complex.

Advanced data processing, analysis, interpretation and modelling provides the means of characterizing the Earth's regional trends and with such a representation; it is possible to predict the Bouguer anomalies and other densities acceptable for models. The knowledge of the free-air anomaly of the Earth enables the gravity anomaly to be determined over a survey area from measurements of the gravitational field strength. The method were applied to real field measurements of Bouguer gravity data over the Sokoto Basin, Nigeria. The working data were corrected for Bouguer reduction density variation using regional-residual separation model. In particular, the major anomalies of the regional and the observed Bouguer gravity field exhibits majorly trending in the E-W, and NW-SE directions adjacent to the main structural fold patterns of (**Figures 9b** and **11**) in the northwestern parts. The anomaly field which is the summary of the regional field was further processed to obtain the residual gravity anomaly (**Figure 13**). The regional models show that the crustal structure in the study area consists of normal continental crust, which is divided into lower and upper by the Conrad of a nearly constant depth. The density effect for the sedimentary formations is necessary valuable and extremely critical to interpret the deeper effect of Sokoto Rima groups. Also, it may occur on the edges, in other gravity filtering enhancements which significantly influence the structural fittings. In order to overcome the edge effect, the modelled length is slightly enlarged outside the limits.

Acknowledgements

I acknowledge the words of encouragement support of my chairman supervisory team for my ongoing Ph.D. Program (Prof. O.K. Likkason) of Physics Program, ATBU, Bauchi, Nigeria as well as members' supervisory team (Prof. A.S. Maigari & Dr. S. Ali) of ATBU Bauchi, Nigeria. Indeed I'm grateful to the support intervention for Petroleum Technology Development Fund, PTDF, Abuja, Nigeria. Finally, I also wish to acknowledge the reference text citation made within context of this work.

Author details

Adamu Abubakar[1*] and Othniel K. Likkason[2]

1 Department of Applied Geophysics, Federal University Birnin Kebbi, Kebbi State, Nigeria

2 Department of Physics, Abubakar Tafawa Balewa University, Bauchi, Nigeria

*Address all correspondence to: adamu.abubakar35@fubk.edu.ng; talk2adamuabubakar35@gmail.com

IntechOpen

References

[1] Kearey P, Brooks M, Hill I. An Introduction to Geophysical Exploration. 3rd ed. Oxford: Blackwell Publishing; 2002

[2] Lowrie W. Fundamental of Geophysics. 2nd ed. Cambridge: Cambridge University Press; 2007. pp. 43-73

[3] Olawale OO, Moroffdeen AA, Oluwatoyin AA. Structural interpretation and depth estimation from aeromagnetic data of Abigi-Ibebu-waterside area of Eastern Dahomey Basin, Southwestern Nigeria. Geofisica International. 2020;**58**(4):29-37

[4] Kayan KR. Potential Theory in Applied Geophysics. Heidelberg, New York: Springer; 2008

[5] Heiskanen WA, Vening Meinesz FA. The Earth and Its Gravity Field. New York: McGraw Hill Book Company; 1958

[6] Blakely RJ. Potential Theory in Gravity and Magnetic Applications. Cambridge: Cambridge University Press; 1996

[7] Blakely RJ. Potential in Gravity and Magnetic Applications. XiX+441. Cambridge, New York, Port-Chester, Melbourne, Sydney: Cambridge University Press; 1995. pp. 307-308. ISBN: 052141508X

[8] Telford WM, Geldert LP, Sheriff RE, Keys DA. Applied Geophysics. Cambridge, London, Sydney: Cambridge University Press; 1976

[9] Talwani M, Ewing M. Rapid computation of gravitational attraction of three dimensional bodies of arbitrary shape. Geophysics. 1961;**25**:203-225

[10] Wilcox LE. Gravity data, advanced processing and gravity anomaly interpretation. In: James D, editor.

Encyclopedia of Solid Earth Geophysics. Dordrecht: Springer; 1989. pp. 601-619. ISBN 9789048187010

[11] Nettleton LL. Regionals, residuals, and structures. Geophysics. 1976;**19**(1):1-22

[12] Ismail AM, Sultan SA, Mohamady MM. Bouguer and total magnetic intensity maps of Sinai Peninsula, Scale 1:500,000. In: Proc. 2nd International Symposium on Geophysics, Tanta. 2001. pp. 111-117

[13] Osazuwa IB. Cascade model for the removal of drift from gravimetric data. Survey Review. 1988;**29**(228):295-303

[14] Adamu A, Likkason OK, Maigari AS, Ali S. Structural mapping insight from gravity data for hydrocarbon accumulation in some parts of the Sokoto Basin Northwestern Nigeria. In: 7th Assembly of Arab Conference on Astronomy and Geophysics (ACAG-7). Cairo, Egypt: National Research Institute for Astronomy and Geophysics (NRIAG); 2021

[15] Osazuwa IB. Method of execution and analysis of observations of the primary gravity network of Nigeria. International Journal of BioChemiPhysics. 1991;**1**(2):30-36

[16] Obaje NG, Aduku M, Yusuf I. The Sokoto Basin of Northwestern Nigeria: A preliminary assessment of hydrocarbon prospectivity. Petroleum Technology Development Journal. 2013;**3**(2):66-80

[17] Kogbe CA. Geology of the upper cretaceous and lower tertiary sediments of the Nigerian sector of the Iullemmeden Basin (West Africa). Geologische Rundschau. 1972;**62**(1):197-211

[18] Kogbe CA. Geology of Southeastern Portion of the Iullemmeden Basin

(Sokoto Basin). Zaria, Nigeria: Bulletin of the Department of Geology, Ahmadu Bello University; 1979. p. 420

[19] Kogbe CA. Cretaceous and tertiary of the Iullemmeden Basin in Nigeria (West Africa). Cretaceous Research. 1981;2:129-186

[20] Obaje NG. Geology and Mineral Resources of Nigeria: Lecture Notes in Earth Sciences. Berlin Heidelberg: Springer; 2009

[21] Griffin WR. Residual gravity in theory and practice. SEG-Geophysics. 1949;14(1):39-56. DOI: 10.1190/1.1437506

[22] Bhattacharyya BK. Some general properties of the potential fields in space and frequency domain: A review. Geoexploration. 1967;5:127-143

[23] Stacey FD. Physics of the Earth. 3rd ed. Brisbane: Brook Field Press; 1992

Section 3

Seismic Forecasting, Seismotectonics and Geodynamic Evolution of the Himalayan Belt

Unveiling the Evolution Journey from Pangea to Present Himalayan Orogeny with Relation to Seismic Hazard Assessment

Tandrila Sarkar, Abhishek Kumar Yadav, Suresh Kannaujiya,
Paresh N.S. Roy and Charan Chaganti

Abstract

The objective is to understand incessant seismic activities in Northwest and Central Himalayan regions. GPS data acquired (2017–2020, Nepal; 2015–2019, Uttarakhand) from 65 GNSS stations are used to generate velocity solutions with respect to International Terrestrial Reference Frame 2014 & Indian fixed reference frame to determine the site's precise position. These velocities are further used to calculate the strain rate and prevailing convergence rate by the respective Triangulation method and Okada's formulation. The estimated mean maximum and minimum principal strain rate are 12.19 nano strain/yr. and − 102.94 nano strain/yr. respectively. And the respective mean shear strain and dilatation are 115.13 nano strain/yr. −90.75 nano strain, which implies that Higher Himalaya observes high compression rate compared to Outer and Lesser Himalayan region. Estimations have also elucidated presence of extensional deformation in the Northwestern part of the Himalayan arc. Accordingly, in Central Himalaya, paleoliquefaction investigations have deciphered turbidites, confirming that the seismic ruptures did not reach the surface during the 2015 Gorkha earthquake. The best-fit locking depth of 14 km and convergence rate of 21 mm/yr. (Nepal) & 18 mm/yr. (Uttarakhand) are obtained. The strain budget analysis indicates that Northwest and Central Himalaya can beckon a megathrust earthquake in the future.

Keywords: triangulation method, Okada formulation, palaeoliquefaction, best-fit estimation, Himalaya

1. Introduction

The collision between the Indian and Eurasian Plate gave birth to the world's youngest orogeny or fold mountain known as Himalaya. The Himalayan orogeny is very young compared to Aravallis and the Appalachian Mountains. During the mid-Permian, a shallow sea known as Tethys was present in the latitudinal area now covered by the Himalaya. In this era, the Pangea had also started splitting up into small landmasses. In this course, the Northern Eurasian landmass or Angara and the Southern Indian landmass or Gondwana deposited a reasonable amount of sediments into the Tethys, reducing the gap between the Eurasian and Indian landmass.

The mountain building process took ~55 Ma, which started in the Upper Cretaceous period and continued until the Middle Miocene. In the last stage, the formation of Siwaliks also took place that continued from Early Miocene to Neogene age [1, 2].

The convergence activity in the Himalayan arc has made it very vulnerable or important to high seismic activity [3], leading to megathrust earthquakes in the brittle part of the crust that has accumulated elastic strain for a prolonged period. Also, the complex rheology of Himalayan wedges and crustal structure is another reason for the seismic hazard potential. Past studies have shown that Uttarakhand and Western Nepal had been the most potential zone for megathrust earthquakes for 200–500 years [4–7]. Therefore, in the past half-century, the Northwest Himalaya was ruptured by various small and large earthquakes. Similarly, in the Central Himalaya, the recent Gorkha earthquake of ~Mw >7 that took place in Kathmandu, Nepal on 25 April 2015 resonated with the 1833 megathrust earthquake, thereby indicating an ongoing convergence rate of ~14–20 mm/yr. (also consistent with the estimated value) as shown in Section no. 6 & 7 in the Himalayan region [8].

A dense network of GNSS stations with ~ > 5 years long period also accomplices in crustal deformation study and help in detecting the slip rate asperity on decollement to unravel the zones/areas with a high risk of seismicity. A detailed study is needed to understand the factors and precursors that trigger megathrust earthquakes. This research work estimated the surface deformation to assess the future seismic hazard potential in Northwest-Central Himalayan region using a close-knit network of 65 GNSS stations (both permanent and campaign).

2. Tectonic setting

Himalaya is located along the southern fringe of Tibetan plateau with a stretch of ~2500 km length and 250–300 km width, is bounded by Nanga Parbhat (Indus gorge) in northwest and Namcha Barwa (Tsangpo gorge) in the Northeast. The morphologic and structural framework of Himalaya is classified as Kashmir-Punjab of length ~ 550 km, Kumaun-Garhwal of length ~ 320 km, Assam segments of length ~ 400 km and Nepal of length ~ 800 km [9]. At the east and west syntaxes, the Tsangpo River separates Himalaya from the Indo-Burmese range in the east, and the Indus River separates Hindukush in the west, respectively. The Himalaya with peaks high resides on compressed, thrusted, folded and 70 km thick (twice the thickness from the normal continental crust) continental crust, witnessed through geophysical surveys (seismic reflection and gravity) [10–12]. MFT lies adjacent to Indo-Gangetic Plain a foreland formed due to convergence between the Indian and Eurasian Plate. The Main Himalayan Thrust acts as a décollement or detachment that dips at a shallow depth. The South Tibetan Detachment, the Main Central Thrust, the Main Boundary Thrust, the Himalayan Frontal Thrust coalesce and meet [13, 14]. The Formation of the Himalaya took place in five distinct phases, which are described in **Table 1** relative to their geology and age modified after [5, 15, 16].

3. GPS velocity measurements within close-knitted GNSS stations

Himalaya has a history of seismicity or earthquakes of very high magnitude that lead to devastation and casualties. To assess the seismic hazard potential, the interseismic strain that is getting accumulated for quite a long period is evaluated with the aid of a close-knitted network of continuous and campaign GNSS stations [5, 6, 17, 18]. These stations are located exquisitely along the Northwest and Central Himalayan stretch spanning the three major thrusts (HFT, MBT and MCT, **Figure 1**).

Evolution Phase	Age of formation	Tectonic/geological process	Geology (metamorphism grade, Fossils)
Trans-Himalayan Uplift	55–35 Ma	Upliftment due to thrusting on South-directed thrust such as Gangdese Fault, Transportation of sediments by Indus and Tsangpo river, rapid cooled and eroded igneous rocks, formation of major basins such as Kargil, Kailash and Lhasa	Granitic and volcanic rock intruded by metamorphic and sedimentary rock of the South Tibetan block
Tethyan Himalaya Uplift	45–35 Ma	Extensive folding, uplift of granite gneiss domes, formation of North Himalayan thrust, 'antecedent' pattern of Himalayan rivers such as Indus, Satluj, Kali, Ganga, Gandaki, and Brahmaputra	Cambrian-Eocene sandstone, shale and limestone (contains fossils which includes cretaceous age ammonites), Miocene Leucogranite, Post early middle Eocene-Biotite grade and lower; Early Miocene and younger-Kyanite-sillimanite grade
Higher Himalaya Uplift	24–17 Ma	Uplift of the metamorphic rocks buried at 20–25 km depth along the MCT, separation of Tethyan Himalaya from Higher Himalaya along STD that is characterized by tectonic extension, gravitational gliding and back-folding of the sediments of Tethyan Himalaya and upliftment and exhumation of the Higher Himalaya, Klippe formation, deposition of sediments in the foothills of Himalaya	10–20 km thick schist and Gneiss, Miocene leucogranite, Central Crystalline Zone, Post-Early middle Eocene-garnet grade
Lesser Himalaya uplift	11–7 Ma	Thrusting across MBT, began of Monsoon, rapid erosion of Himalaya, excessive rate of sedimentation in the Siwalik basin,	Quartzite, marble, slate, schist and gneiss, Eocene-Chlorite grade and lower
Neotectonics	2.6–0 Ma	Uplift of Siwalik range along HFT, rapid erosion, deep and narrow gorges, abundant of coarse-grained fluvial sediments	Sandstone, mudstone (abundant of mammal fossils), Unmetamorphosed

Table 1.
Evolution phase of the Himalaya with associated tectonic processes and geology.

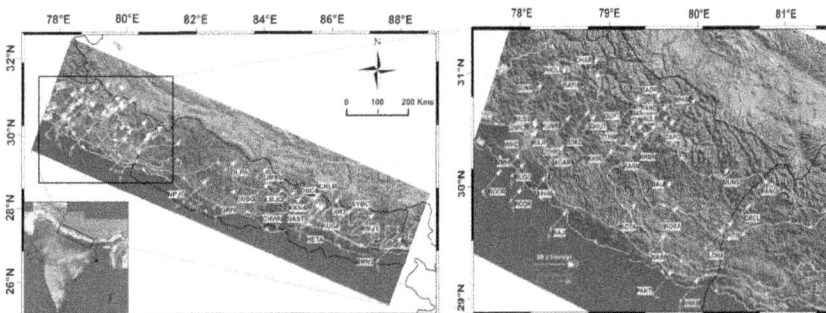

Figure 1.
A schematic diagram illustrating the study area along the northwest-central Himalayan stretch and the locations of the 65 GNSS stations. The square on the left figure is enlarged on the right highlighting the northwest Himalaya or Uttarakhand region along with the GNSS stations that are located in this realm. Note the white vectors indicate the velocity with respective to ITRF2014 and the white circles on it defines the error ellipse. While the red vectors represent the velocities derived with respect to India fixed reference frame (Jade et al. [19]) and the red circles on it define the seismicity of varying magnitude. The yellow lines feature the three Main thrusts of Himalaya; Himalayan frontal thrust (HFT), Main boundary thrust (MBT) and Main central thrust (MCT).

The Continuous GNSS stations or CORS (Continuously Operating Reference Stations) are well equipped with robust & firmly constructed receivers from Trimble and Leica, Choke-ring geodetic antenna, <10° elevation angle and the time taken for data recording is at 15 s sampling interval. Consequently, the campaign or episodic GNSS stations are equipped with a Zephyr antenna. The data is recorded for 3–4 days repeatedly at regular intervals so that the hydrological mass distribution and seasonal variation effect in this region diminish accordingly. The data is acquired from 65 stations for 2015–2019 (Uttarakhand region) and 2017–2020 (Nepal region). GPS data acquired is processed in GAMIT/GLOBK (vs. 10.71) software along with nearby IGS (International GNSS Services) stations to calculate the precise position of these 65 stations [20]. It takes two steps to complete the GPS data processing. In the first step, the daily relative position of an individual station/site is estimated with respect to clock errors and orbital parameters of the satellite, along with error predicted due to ocean-tidal effect, ionospheric electron content and atmospheric water vapor. The Finite Element Solution (FES) 2012 and International Earth Rotation system (IERS) 2010 models are employed to reduce the respective ocean tidal and Earth tidal effects. While the Global Pressure (GPT) model and Global Mapping Function (GMF) correct the tropospheric delay, which results due to wet and dry water mass [21]. The final or the second step of this process uses GLOBK modules to analyze the loosely constrained solution for obtaining the site velocity with respect to International Terrestrial Reference Frame 2008 (ITRF08) and ITRF14 Uttarakhand and Nepal region, respectively. The velocities obtained from ITRF08 is converted to ITRF14 using the HTDP software (geodesy.noaa.gov/TOOLS/Htdp/Htdp.shtml). Also, the velocities for each station is calculated with respect to fixed Indian reference Frame/Euler Pole as suggested by [19] (**Table 2, Figure 1**). The velocities

SITE	LONG	LAT	East_vel	Error (east)	North_vel	Error (north)	Fixed Indian (North)	Fixed Indian (East)
BADR	79.49	30.74	27.4	1.2	21.3	1	−10.7	−3.6
BUGG	77.89	30.08	35.9	1.1	30.6	0.9	−2.5	0.3
DEHR×	78.04	30.34	34	0.3	35	0.4	0.1	0.1
GNFC×	78.05	30.45	33.3	0.4	33	0.5	−1.9	−0.5
HHIM	78.16	29.95	31.1	1	34.5	0.8	0.1	−1.1
KARN	79.21	30.26	28.9	1.2	30.1	0.9	−5.1	−5.4
MAND×	79.27	30.45	32.9	0.7	26.1	0.6	−9.1	−1.2
MHIM	79.89	30.68	27.7	1.2	23.7	1.3	−11.6	−6.5
MMHI	77.88	30.18	31.3	0.8	32	0.8	−1.9	0.1
MUSS	78.09	30.45	31.1	0.9	30.4	0.8	−2.8	−1
PAND	79.55	30.63	33.5	1.2	24.1	1	−11.2	−0.6
PANT×	79.48	29.01	36.2	0.5	35	0.4	−0.3	1.1
PIPA	79.43	30.44	29.4	0.9	28.6	1.1	−6.6	−4.8
RAJP	78.09	30.38	27.4	1.2	32.1	0.8	−0.6	−6.2
ROOR	77.88	29.85	34.8	1.1	31	1	−2.6	1
SAHA	77.56	29.93	34.9	1	33.8	1.2	−0.9	0.9
TAPO	79.64	30.49	27.8	1.1	25.6	1.2	−9.7	−6.5
THIM	79.20	30.49	29.5	0.9	29.6	1	−5.6	−4.6

SITE	LONG	LAT	East_vel	Error (east)	North_vel	Error (north)	Fixed Indian (North)	Fixed Indian (East)
BRN2	87.27	26.51	39.06	0.05	34.97	0.06	−1.94	0.16
CHLM	85.31	28.20	31.97	0.04	13.89	0.05	−22.67	−5.42
CHWN	84.38	27.66	36.86	0.14	32.6	0.14	−3.78	−0.56
DLPA	82.81	28.98	35.42	0.1	24.7	0.12	−11.35	−0.77
DNC4	85.24	28.07	28.18	0.21	17.67	0.24	−18.88	−9.27
DNSG	83.76	28.34	35.85	0.04	27.46	0.05	−8.80	−1.00
GRHI	82.49	27.95	38.31	0.37	29.88	0.4	−6.10	1.60
HETA	85.05	27.41	37.39	0.06	33.66	0.07	−2.85	−0.37
JIR2	86.18	27.65	35.5	0.04	24.13	0.05	−12.59	−2.46
JMSM	83.74	28.80	35.25	0.04	24.33	0.06	−11.92	−1.32
KKN4	85.27	27.80	34.09	4.05	29.53	4.83	−7.03	−3.52
KUGE	85.53	27.61	35.95	0.13	33.57	0.13	−3.04	−1.84
LHAZ	91.10	29.65	46.38	0.04	15.63	0.05	−21.84	7.96
LMJG	84.57	28.17	34.49	0.04	25.86	0.05	−10.56	−2.70
NAST	85.32	27.65	32.44	0.07	34.25	0.08	−2.32	−5.27
NPGJ	81.59	28.11	38.34	0.05	34.49	0.07	−1.29	1.99
RMJT	86.55	27.30	38.21	0.04	29.38	0.05	−7.41	−0.05
SYBC	86.71	27.81	38.35	0.04	21.72	0.04	−15.10	0.31
BDRI	79.49	30.74	27.56	0.3	23.67	0.1	−5.3	12.6
MALA	79.88	30.68	30.1	0.1	22.9	0.2	−11.6	−6.5
DHAR	78.78	31.04	29.4	0.1	24.2	0.1	−13	−1.4
MUNS	80.25	30.06	30.7	0.9	25.9	1.8	−11.6	−4.9
DRCL	80.5	29.73	30.3	0.3	26.5	0.5	6.3	9.7
PANG	80.69	29.97	34.1	0.8	21.5	0.9	−10.1	−5
PHOL	78.42	30.95	29.9	0.4	27.7	0.8	−8.7	−4.6
HRMN	79.33	30.38	30.6	0.2	27.3	0.2	−8	−5.3
BAIJ	79.60	29.93	31	1.3	27.5	1.7	−7.5	−3.6
MUN2	80.24	30.06	31.13	0.6	27.96	0.3	−4.3	−7.1
AULI	79.56	30.53	33.1	0.4	27.7	0.5	−7	−4.2
GHUT	78.74	30.53	30.96	0.2	30.88	0.1	−6.6	−4.5
GUTU	78.74	30.53	32.1	0.1	31.2	0.2	−6.7	−1.8
BDKD	78.66	30.55	32.2	0.2	31.1	0.2	−5.5	−3.2
PITH	80.28	29.57	32.1	0.5	31.4	0.5	−4.2	−3
SUNR	78.13	30.79	31.4	0.1	32.3	0.3	−2.7	−3.5
PITH2	80.28	29.57	31.55	0.5	32.77	0.3	−3.3	−2.6
AGAR	78.33	30.21	34.2	0.1	32.8	0.2	−3.2	−2.5
WIHD	78.01	30.32	32.67	0.1	34.59	0.2	−2	−2.9
KUNR	78.40	30.46	33.5	0.5	33.8	0.4	−2.6	0.5
KHIR	78.88	30.16	33.8	0.1	33.6	0.1	0.7	−2.5
KORA	79.59	29.64	35.8	2.1	31.8	0.9	−0.3	−1.7

SITE	LONG	LAT	East_vel	Error (east)	North_vel	Error (north)	Fixed Indian (North)	Fixed Indian (East)
NAIN	79.45	29.38	33.7	0.2	34.2	0.3	−0.2	−1.6
NAJI	78.33	29.59	34.8	0.7	33.1	0.7	−1.5	−0.5
LOHA	80.08	29.42	34	0.3	34.3	0.2	−0.8	−1.1
NNKM	79.80	28.94	33.2	0.7	35.2	0.7	−1.3	−0.2
KOTA	79.09	29.64	34.6	0.1	34.3	0.2	−0.3	−1.2
RATH	78.60	30.81	47.1	0.7	29.1	0.2	−0.2	−0.6
GUPT	79.07	30.53	44.4	0.6	40.7	0.5	−0.2	−0.1

The starred (×) stations indicate the permanent ones.

Table 2.
Estimated velocities from both continuous and campaign stations with respect to ITRF 2014 and Indian fixed reference frame [19].

estimated with respect to ITRF14 vary from 34.70 ± 1.5 (BADR) to 52.42 ± 0.07 (BRN2), thereby indicating crustal shortening rate in the Higher Himalaya is very high compared to the stations that are located in the Outer Himalayan region. Accordingly, the velocities obtained from India Plate fixed reference frame, vary from 23.3 (CHLM) to 0.1 (DEHR). Such variation signifies that the surface deformation in the Himalayan stretch occurs due to the thrusting of the Indian Plate below the Eurasian Plate.

4. Strain rate estimation on a localized scale using the triangulation method

The estimated site velocities depend on the deformation rate and the translation motion [22]. The translation motion does not depend on the position in a localized zone and has a similar direction and magnitude for a particular reference frame. Whereas the deformation rate indicates a change in the displacement (Δu_i) or position of the site secularly, which is explained in 2D by the following equation.

$$\Delta u_i = L_{ij} \Delta x_j \ \text{ for } i,j = x,y \tag{1}$$

where Δx_j is the change in the site position and L_{ij} is the displacement gradient tensor. The L_{ij} or displacement gradient tensor is decomposed into asymmetric and symmetric parts (Eq. 2), thereby representing the rotation rate and the infinitesimal strain rate.

$$L_{ij} = \varepsilon_{ij} + \omega_{ij} = \frac{1}{2}\left(\frac{\partial u_i}{\partial x_j} + \frac{\partial u_j}{\partial x_i}\right) + \frac{1}{2}\left(\frac{\partial u_i}{\partial x_j} - \frac{\partial u_j}{\partial x_i}\right) \tag{2}$$

Within a close clustered network of GNSS stations that are distributed along the Himalayan stretch, the estimation of strain rate (on a local scale) and the rate of rotation can be carried with the aid of the Triangulation method [23]. In this method, the horizontal field velocity ($\dot{u}_x, \dot{u}_y$) determines the deformation rate at the centroid of the triangle at three nonlinear stations. Each of the horizontal velocity component is expressed in terms of deformation rate ($\dot{\varepsilon}_{xx}, \dot{\varepsilon}_{xy}, \dot{\varepsilon}_{yx}, \dot{\varepsilon}_{yy}$) and translation motion (t_x, t_y) with the recognized initial site location (x_o, y_o). So, the velocity field at three different stations gives six equations (i.e. 2 for each site) alongwith six unknown, and it is expressed by the following equation.

$$
\begin{pmatrix}
{}^1\dot{u}_x \\
{}^1\dot{u}_y \\
{}^2\dot{u}_x \\
{}^2\dot{u}_y \\
{}^3\dot{u}_x \\
{}^3\dot{u}_y
\end{pmatrix}
=
\begin{pmatrix}
1 & 0 & {}^1x_o & {}^1y_o & 0 & 0 \\
0 & 1 & 0 & 0 & {}^1x_o & {}^1y_o \\
1 & 0 & {}^2x_o & {}^2y_o & 0 & 0 \\
0 & 1 & 0 & 0 & {}^2x_o & {}^2y_o \\
1 & 0 & {}^3x_o & {}^3y_o & 0 & 0 \\
0 & 1 & 0 & 0 & {}^3x_o & {}^3y_o
\end{pmatrix}
\begin{pmatrix}
t_x \\
t_y \\
\dot{\varepsilon}_{xx} \\
\dot{\varepsilon}_{xy} \\
\dot{\varepsilon}_{yx} \\
\dot{\varepsilon}_{yy}
\end{pmatrix}
\tag{3}
$$

The above equation and its solution for three non-collinear GPS stations can be written in the form of

$$\dot{u} = Gm \tag{4}$$

$$m = G^{-1}\dot{u} \tag{5}$$

where $\dot{u}$ is a 1×6 column matrix of known instantaneous displacement/velocity vectors at three stations, m is a 1×6 column matrix of unknown translation vector and the deformation gradient tensor, and G is a square matrix of 6×6 order, and G^{-1} is its inverse matrix which includes zeros, ones and the location vector coordinates of the three GNSS stations.

In a triangulation network, the translation gradient, rotational gradient, maximum shear strain rate and dilatation is estimated using the derived model parameters [24].

The translation gradient is calculated as:

$$Speed = \sqrt{t_x^2 + t_y^2} \tag{6}$$

The rotation rate is expressed as:

$$
\dot{\omega} =
\begin{pmatrix}
0 & \dfrac{(\dot{\varepsilon}_{xy} - \dot{\varepsilon}_{yx})}{2} \\
\dfrac{(\dot{\varepsilon}_{yx} - \dot{\varepsilon}_{xy})}{2} & 0
\end{pmatrix}
\tag{7}
$$

2D Lagrangian strain rate tensor described as;

$$
\dot{\varepsilon}_{ij} =
\begin{pmatrix}
\dot{\varepsilon}_{xx} & \dfrac{\dot{\varepsilon}_{xy} + \dot{\varepsilon}_{yx}}{2} \\
\dfrac{\dot{\varepsilon}_{yx} + \dot{\varepsilon}_{xy}}{2} & \dot{\varepsilon}_{yy}
\end{pmatrix}
\tag{8}
$$

The magnitude and orientation of maximum (e_1) and minimum (e_2) principal strain rate tensor. Also, compression and extension in a regime are determined by the respective negative and positive values of principal strain rates. And the maximum infinitesimal shear strain (γ_{max}) is expressed by the following equation.

$$\gamma_{max} = e_1 - e_2 = 2\sqrt{\left(\frac{\varepsilon_{xx} - \varepsilon_{yy}}{2}\right)^2 + (\varepsilon_{xy})^2} \tag{9}$$

The first invariant of the 2D strain tensor is the areal strain or dilatation ((ellipse area-circle area)/circle area) and is explained by the following equation.

$$\text{Area strain rate/Dilatation} = e_1 + e_2 \qquad (10)$$

5. Inferences from localized strain rate estimation

The strain rate and rotation rate are estimated at the centroid of 89 triangular zones of the 65 sites forming an angle of 30° or more (**Table A1**). The maximum strain rate varies from -47.90 to 195.22 nano strain/yr. with an azimuth variation from 23.21°N to 161.02° N. Similarly, the minimum strain rate varies from -333.606 to -8.9 nano strain/yr. with an azimuth variation from 178.55 °N to 0.07 °N. The value of maximum shear strain lies between 13.47 to 301.51 nano strain/yr., and the angle of rotation range from -9.9 E-0.7 to 9.12E-0.7 (°/yr). Along with this, the change in dilatation or areal strain differs between 136.07 to -381.405 nano strain. From these observations, it can be stated that the stations (mainly station number 71–86) located in the Higher Himalaya (South East) of the Nepal region experiences a high rate of compression (**Figure 2, Table A1**), clockwise motion with a mean shear value of 115.13 nano strain/yr. (**Figure 4**) and negative dilatation (highlighted by blue to yellow color, **Figure 3**). On proceeding towards the Northwest Himalaya, the stations located there observe a low compression rate (especially along the MBT boundary). At the same time, the cluster of stations (mainly station number 36–50) located in the Higher Himalaya of Uttarakhand region (Northwest)

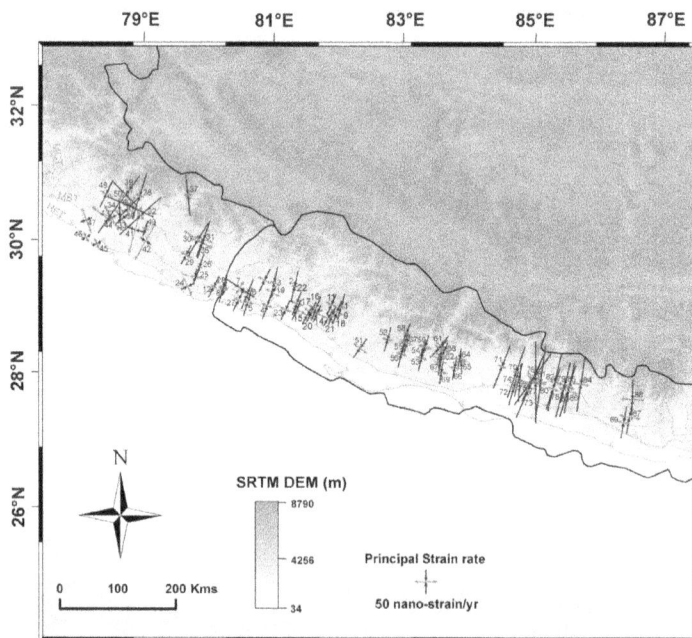

Figure 2.
*The strain rate analysis using the triangulation approach. The red and blue vectors indicate maximum and minimum principal strain rate respectively. Note that the blue vectors are highly dominant towards the south eastern part of Nepal region indicating a high rate of compression. While in the northwest part of Uttarakhand region the red vectors are much dominant thus indicating an extension. The maximum and minimum values estimated are tabulated in **Table A1**.*

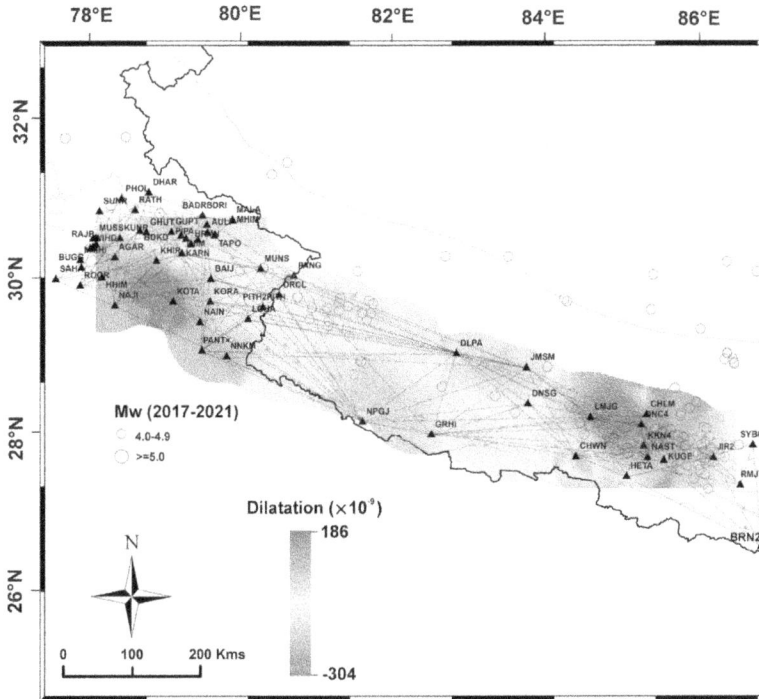

Figure 3.
*The colored patches show the dilatation at the Centre of the respective triangles as described in **Table A1**. The prominent red patch in northwest of the Uttarakhand region defines extension (with reference to **Figures 2** and **4**, here the rate of extension is high). And the prominent blue patch in the Southeastern part of Nepal region indicates high compression rate.*

reflects positive dilatation (highlighted by dark red color, **Figure 3**), anticlockwise movement (**Figure 4**) and positive strain rate values (**Figure 2**) or extension.

The second invariant of the strain rate field, where the second invariant is defined as;

$$\dot{\varepsilon}^2{}_{\varphi\varphi} + \dot{\varepsilon}^2{}_{\theta\theta} + 2\dot{\varepsilon}^2{}_{\varphi\theta} \tag{11}$$

where $\dot{\varepsilon}_{\varphi\varphi}$, $\dot{\varepsilon}_{\theta\theta}$ and $\dot{\varepsilon}_{\varphi\theta}$ are the horizontal components of the strain rate tensor. When the second invariant is estimated, the whole tectonic regime experiences moderate to high areal dilatation (color varies from yellow to orange to red, **Figure 5**), leaving out certain portions of the Eastern or NNE region (blue to yellow color, **Figure 5**). The 2D second invariant to the strain rate tensor has been estimated independently from the principal components of strain (**Figure 5**) to establish a relation between the rate of deformation, seismicity and the interseismic strain accumulating for a prolonged period [25, 26]. The map of the second invariant also focuses on the spatial distribution of earthquakes (**Figure 5**).

In an orogen of convergence, observation of such paradox situation becomes quite inquisitive [27]. Past studies have shown that prominent extensional defor- mation exists in an orogen of convergence like Andean cordillera, Scandinavian Caledonides etc. [28, 29]. In the Himalayan realm, intra-continental tectonic fea- tures, prominent folds, and several other deformational geological structures indi- cate a compressional tectonic area. But in recent studies, it has been observed that normal faults and many other extensional geological structures are also distributed in different Himalayan sectors [30], but these are not consistent with ongoing

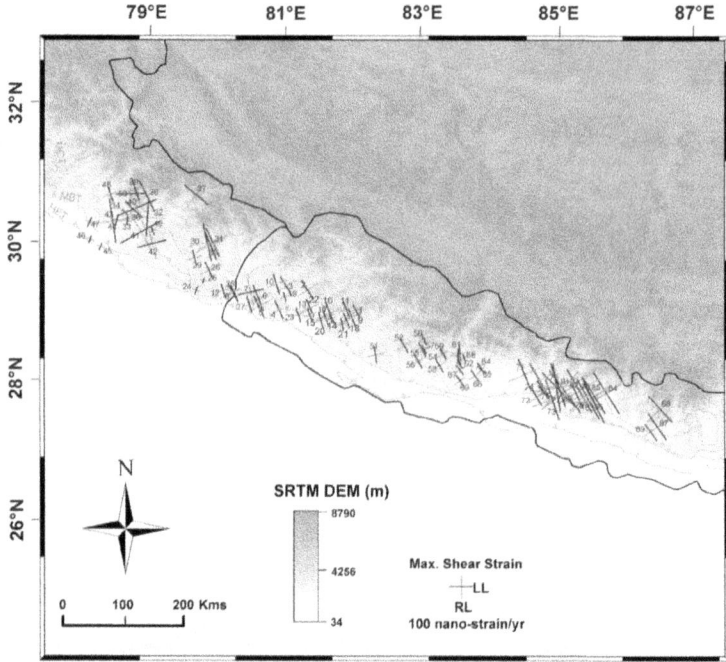

Figure 4.
The maximum shear strain rate is estimated by the triangulation approach. Note the green and blue lines represent right lateral (RL) and left lateral (LL) respectively. The respective estimates and direction of the angular rotation is described in **Table A1.**

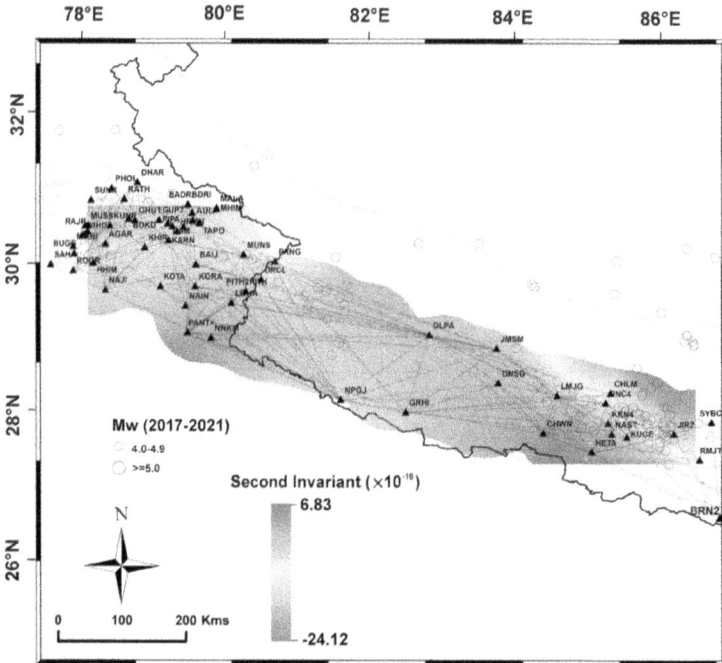

Figure 5.
A 2D second invariant map estimated by the triangulation approach showing the spatial distribution of seismicity. Note the hollow red circles are the seismic events ranging in magnitude from 4 to >5.

seismicity and thrusting in the southward direction. Also, the presence of large scale lineament, shifting and tilting of river, subsidence of older rocks, upliftment of river terraces & active thrusts and normal faults overriding the young Holocene sediments demonstrates that extensional and compressional features co-exist with each other in Himalaya [31, 32].

6. Estimation of ongoing convergence rate and its future implications

The rate of convergence along the active plate boundaries determines the seismic potential and the rate of seismicity occurring today. Previous studies have stated that currently, MFT is the most active fault because here, the crustal short-ening is taking place due to the plate movement and evidence of GPS vectors [33]. The site velocities (obtained with respect to ITRF) are distributed along the strike direction (fault parallel or strike slip fault) (**Figures 6** and **7**) and across the strike direction (fault normal, or oblique fault) (**Figures 8** and **9**), and the structural trend of MFT is taken as 303° and 288° in the Nepal & Uttarakhand region. It is assumed that the fault normal velocity as dip slip and fault parallel velocity as strike slip on the plate interface, 'i.e.' MHT (Main Himalayan Thrust). In fault normal velocity, the uncertainty is calculated by applying the following formula

$$\sigma_{horz} = \sqrt{\frac{\sigma_E^2 \sigma_N^2}{\sigma_N^2 \cos^2\theta + \sigma_E^2 \sin^2\theta}} \tag{12}$$

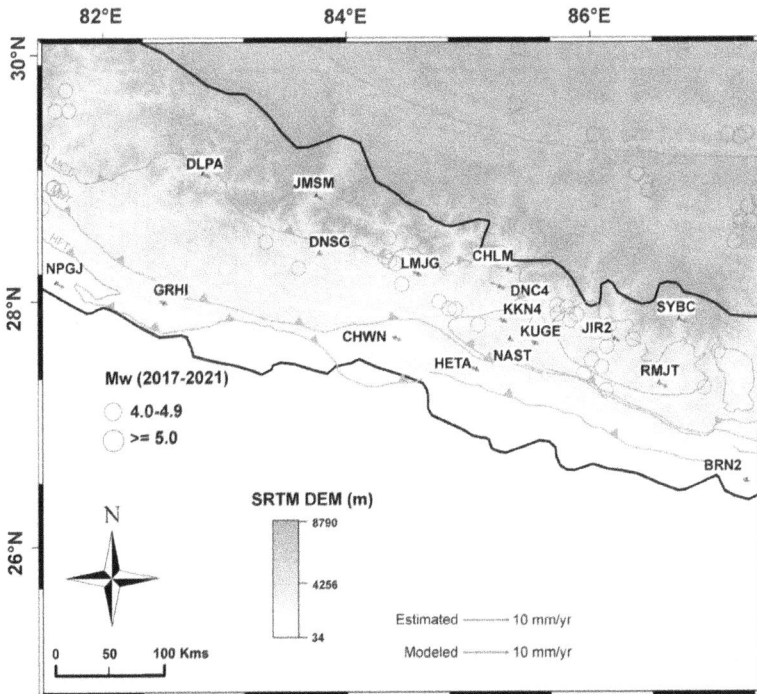

Figure 6.
The rate of ongoing convergence in the Nepal region is estimated along the arc parallel or fault parallel direction. Note the green arrows are the estimated convergence rate while the red arrows define modeled convergence rate. The hollow blue circles indicate seismic events (2017–2021) of magnitude varying from 4 to >5.

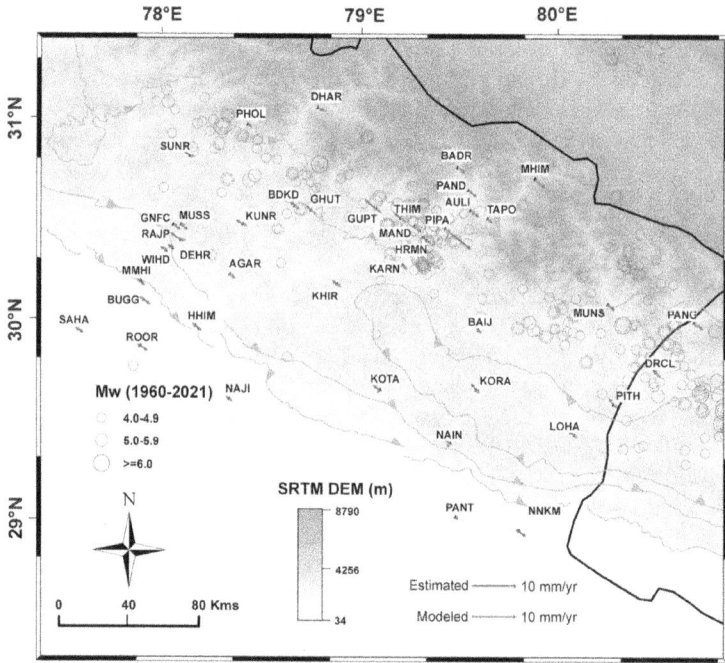

Figure 7.
The rate of ongoing convergence in the Uttarakhand region is estimated along the arc parallel or fault parallel direction. Note the green arrows are the estimated convergence rate while the red arrows define modeled convergence rate. The hollow blue circles indicate seismic events (1960–2021) of magnitude varying from 4 to >6.

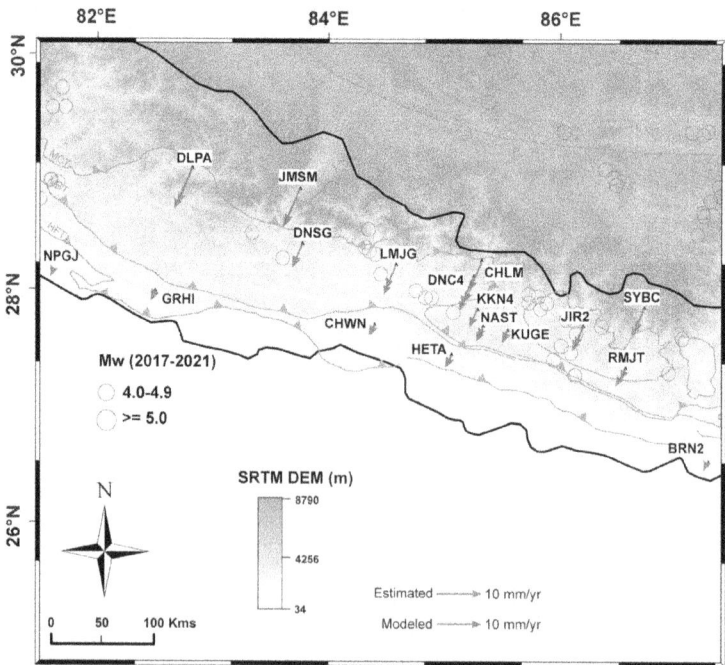

Figure 8.
The rate of ongoing convergence in the Nepal region is estimated across the arc normal or fault normal direction. Note the green arrows are the estimated convergence rate while the red arrows define modeled convergence rate. The hollow blue circles indicate seismic events (2017–2021) of magnitude varying from 4 to >5.

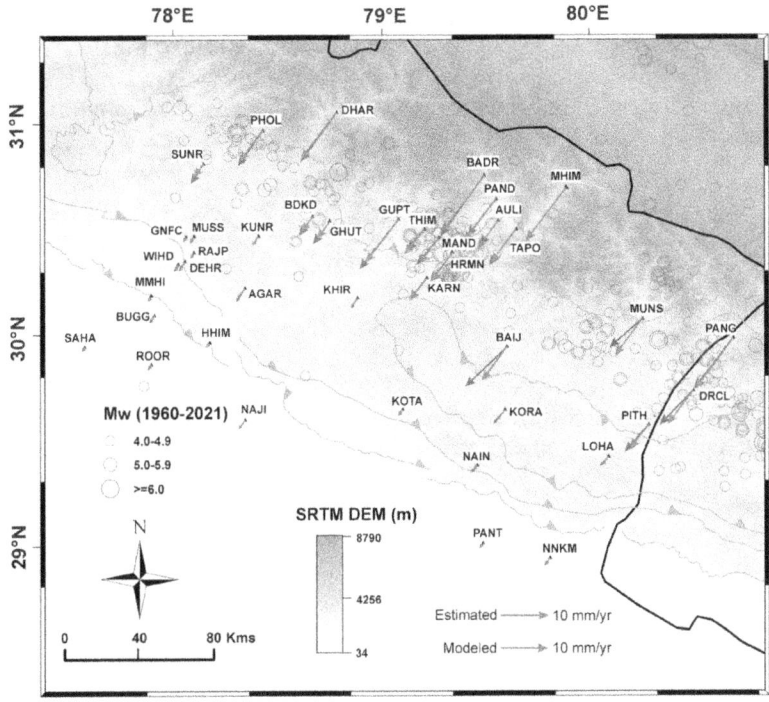

Figure 9.
The rate of ongoing convergence in the Uttarakhand region is estimated across the arc normal or fault normal direction. Note the green arrows are the estimated convergence rate while the red arrows define modeled convergence rate. The hollow blue circles indicate seismic events (1960–2021) of magnitude varying from 4 to >6.

where σ_E is the standard deviation of the east velocity, σ_N is the standard deviation of the north velocity, and θ is the bearing of the convergence vector, counterclockwise from east. At the site (GNSS stations) location, the subsurface deformation generally depends on several fault parameters: rake, dip, strike, fault length, width, etc. This is due to finite subsurface dislocation, and it can be explained by the back slip approach as expressed by [34] in the following expression

$$u = Gs \tag{13}$$

where u is the surface deformation vectors, s is the vector of parameters, and G is the Green's functions matrix computed using the semi-analytical formulation published by [35].

The Okada dip slip dislocation approach [35] estimates the surface deformation at each GNSS site due to a finite rectangular dip slip dislocation on MHT for an elastic half-space. In this approach, it is presumed that the frontal portion of MHT is completely locked to some extent and aseismic creeping occurs with a uniform slip rate at downdip zone of MHT. After that, reduced chi-square uncertainty is calculated because a grid search analysis was required to estimate the best fit value of slip rate and locking depth. The expression used for calculating the chi-square uncertainty is as follows;

$$\chi_v^2 = \frac{1}{(n-P)^2} \sum_{i=1}^{n} \left(\frac{r_i}{\sqrt{f}\sigma_i}\right)^2 \tag{14}$$

where n is the number of observations, P is the number of free parameters, r_i is the residual between observed and calculated velocity, σ_i is the formal data uncertainty, and f is the data scaling factor. The scaling factors helps to avoid the inclusion of any additional uncertainty and prevent influencing various other types of data. Also, the formal GPS velocity uncertainty estimates are generally underestimated by a factor of 2 to 5 [36].

7. Inferences from ongoing plate convergence and seismicity assesment

The plate convergence rate (fault normal) or the elastic strain accumulated in the brittle frontal part of MHT is 21 mm/yr. and 18 mm/yr. for Nepal and the Uttarakhand region, respectively, with 14 km locked depth and dip angle 10° (Nepal) and 7° (Uttarakhand) (**Figures 10** and **11**). But the fault parallel motion is relatively small (3–4 mm/yr) and varies inconsistently (**Figures 6** and **7**), so the fault normal motion is only considered. The slip deficit in the Uttarakhand region is calculated as 3.9 m, indicating that a significant amount of strain had accumulated since the last magnitude earthquake (1803 earthquake) [37]. And the estimated slip deficit corresponds to seismic moment $\sim 4.9 \times 10^{21}$ Nm and $M_w \sim 8.3$ by taking fault length 412 km and assuming its width as 111 Km. Therefore, the strain budget analysis in the Uttarakhand region signifies that this area bears the potential to produce a megathrust earthquake, and the results are consistent with [6, 16, 38, 39]. While the Nepal region was struck by a recent high magnitude (Mw >7) Gorkha earthquake that immensely disturbed the tectonic activity of this region. It has been observed that the 2015 seismicity took place along the planar MHT and also gives good information about ground motion frequencies, the direction of rupture etc. It is also presumed that earthquakes with $M_w > 8$ can rupture the surface devastatingly. But the earthquakes with M_w between 7 and 8 rupture mainly within depth and rarely reach the surface. Also, Paleoliquefaction investigations have deciphered turbidites that formed due to 2015 seismic activity from the Rara Lake, which presents alternative evidence for ruptures that did not reach the surface [40–44]. Therefore, it can be concluded that interseismic strain is building up within the Nepal Himalayan region to beckon the next calamitous earthquake in the near future.

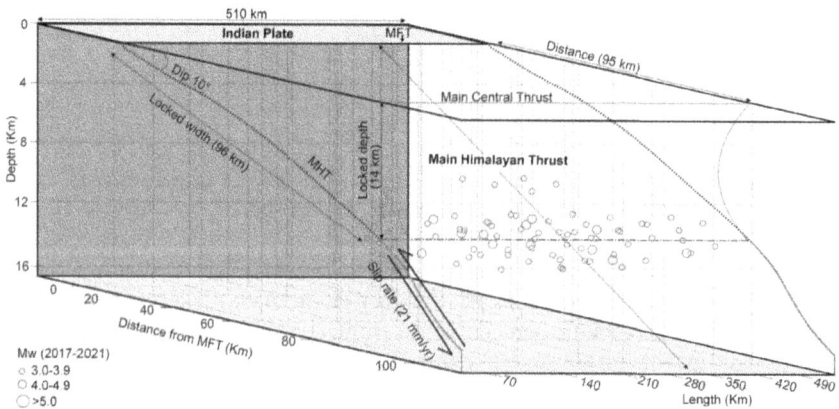

Figure 10.
A 3D reference model showing the cross section of the Himalayan geometry in Nepal region. Note the locked width is 96 km, locked depth is 14 km, dip angle 10° and slip rate estimated is 21 mm/yr. the hollow blue circles represent seismic events from 2017 to 2020 ranging between 3 to >5. Also, the concentration of seismicity is high near the Main central thrust, thereby indicating hypocentre of these earthquakes falls in and around the higher Himalaya.

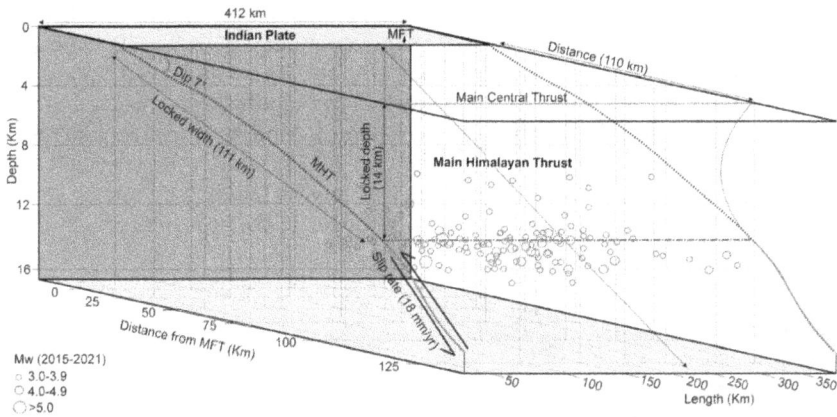

Figure 11.
A 3D reference model showing the cross section of the Himalayan geometry in Uttarakhand region. Note the locked width is 111 km, locked depth is 14 km, dip angle 7° and slip rate estimated is 18 mm/yr. the hollow blue circles represent seismic events from 2015 to 2021 ranging between 3 to >5. Also, the concentration of seismicity is high near the Main central thrust, thereby indicating hypocentre of these earthquakes falls in and around the higher Himalaya.

3D reference model (for both Nepal and the Uttarakhand region) illustrating the predominant occurrence of earthquakes ranging from low to high magnitude is developed using the parameters that have been applied in Okada's formulation. This model discusses the evolution of slip rate spatiotemporally and the related seismic events that happened in and around the Main Central Thrust fault, thereby indicating that the locations of hypocentre fall in the Higher Himalayan region (**Figures 10 and 11**) [41, 45, 46]. But previous studies have shown that many historic megathrust earthquakes have ruptured partially and did not reach the surface [7, 47]. Therefore, through magnitude moment calculations and 3D reference model, it can be presumed that a megathrust event similar to the Gorkha earthquake or even more catastrophic than that will occur instantaneously.

8. Conclusion

The surface deformation rate is measured along the Northwest -Central Himalayan belt using 65 GNSS stations located close to the wide spatial distribution. GPS data acquired (2015–2019, Uttarakhand and 2017–2020, Nepal) from these GNSS stations are used to estimate the velocity solution and determine the precise position of each site with respect to ITRF2014 and Indian fixed reference frame as suggested by Jade et al. [19]. These estimated velocities are then used to calculate the strain rate and prevailing convergence rate of the Himalayan stretch using the Triangulation method and Okada formulation. From the triangulation approach, the mean observed values of maximum principal strain are 12.19 nano strain/yr., minimum principal strain is −102.94 nano strain/yr., maximum shear strain is 115.13 nano strain/yr. and dilatation of −90.75 nano strain. Although the velocity solution from ITRF2014 implies an increase in convergence rate towards the Higher Himalaya, but results from triangulation have also featured comparatively high extensional deformation in the Northwest (Uttarakhand region) than in the Nepal region. Accordingly, using the grid search analysis, the best fit convergence calculated is 21 mm/yr. and 18 mm/yr. in Nepal and Uttarakhand regions. The locking depth is 14 km in both the regions with dip angle 10° (Nepal) and 7° (Uttarakhand).

Considering the last megathrust earthquake in the Uttarakhand region (1803 earthquake), the slip deficit estimated is ~3.9 m corresponding to ~4.9×10^{21} Nm seismic moment and $M_w > 8$. Whereas in Nepal, after the 2015 Gorkha earthquake (M_w 7.3), the tectonic regime is disturbed greatly and has significantly impacted the infrastructure and the GNSS stations present here. Secondary evidences like turbidites from palaeoseismic investigation show that 2015 seismic activity has not ruptured the surface to a megathrust front. This implies building up interseismic stress waiting to release into a calamitous earthquake in the near future.

Acknowledgements

The authors are grateful to the Director of the Indian Institute of Remote Sensing for his consistent encouragement in carrying out this study. We acknowledge those scholars who have assisted us in the field surveys and those who have helped directly and indirectly in this work. We are thankful to Dr. Robert W. King (MIT) for providing GAMIT/GLOBK 10.70 software for processing GPS data. The authors heartily acknowledges the Editor-in-chief and Reviewer for helping in exemplifying their work exquisitely.

A. Appendix

S.NO.	Rotation ± uncertainty (degrees/yr)	Max horizontal extension (e1H) (nano-strain)	Azimuth of S1H (degrees)	Min horizontal extension (e2H) (nano-strain)	Azimuth of S2H (degrees)	Max shear strain (nano-strain)	Area strain (nano-strain)	Long (°)	Lat (°)
1	5.36E-08	0.755184	106.9466	-78.8757	16.94656	79.63087	-78.1205	82.01544	28.97287
2	-4.04E-07	2.578359	103.6807	-99.4337	13.68069	102.012	-96.8553	81.31119	29.27868
3	-4.47E-07	3.897905	104.6112	-97.7991	14.6112	101.6971	-93.9012	81.00014	29.33331
4	-8.06E-07	12.2127	110.9687	-87.5555	20.96874	99.76821	-75.3428	80.89607	28.98761
5	-6.17E-07	11.00359	108.7238	-91.4848	18.72384	102.4884	-80.4813	80.59397	29.04022
6	-1.21E-06	23.8691	108.2314	-99.8852	18.23138	123.7543	-76.0161	80.58626	29.16166
7	-1.21E-06	17.19301	111.0656	-105.442	21.06559	122.6353	-88.2493	80.46837	29.25081
8	-4.80E-07	9.787359	107.9763	-88.2259	17.97631	98.01327	-78.4386	80.21492	29.23759
9	5.15E-07	17.46843	114.1938	-75.7052	24.1938	93.17367	-58.2368	81.94994	28.89188
10	3.83E-07	17.42987	117.4941	-77.7848	27.49407	95.21467	-60.3549	80.85491	29.36348
11	2.50E-07	13.2956	114.4983	-70.6135	24.49831	83.90915	-57.3179	81.87103	29.00414
12	5.06E-07	6.954867	118.3468	-72.1839	28.34675	79.13882	-65.2291	80.06962	29.26691
13	8.85E-08	15.74241	111.9557	-76.4613	21.9557	92.20373	-60.7189	81.34453	29.01988
14	-2.13E-07	5.76327	115.0188	-61.9755	25.01882	67.73874	-56.2122	81.6495	28.86617
15	-3.89E-07	6.863233	114.4877	-70.1774	24.4877	77.04068	-63.3142	81.33987	28.92157
16	2.63E-07	22.81866	112.4271	-77.6835	22.42707	100.5022	-54.8649	81.64041	28.94797
17	-1.50E-07	23.51346	116.8895	-70.4583	26.88947	93.97175	-46.9448	81.56862	28.89528
18	1.99E-07	19.41619	118.9625	-63.8279	28.96255	83.24404	-44.4117	81.87805	28.83936

S.NO.	Rotation ± uncertainty (degrees/yr)	Max horizontal extension (e1H) (nano-strain)	Azimuth of S1H (degrees)	Min horizontal extension (e2H) (nano-strain)	Azimuth of S2H (degrees)	Max shear strain (nano-strain)	Area strain (nano-strain)	Long (°)	Lat (°)
19	-3.99E-07	22.79689	117.718	-45.9614	27.71804	68.75828	-23.1645	80.9715	29.17439
20	-4.13E-07	20.65258	119.0297	-67.1152	29.02969	87.76776	-46.4626	81.50303	28.84527
21	-6.32E-08	18.17555	121.5551	-57.198	31.5551	75.3736	-39.0225	81.81234	28.7895
22	8.85E-08	15.74241	111.9957	-76.4613	21.9557	92.20373	-60.7189	81.34453	29.01988
23	-3.27E-07	12.22406	115.5537	-69.9911	25.5537	82.21512	-57.767	81.17633	28.92427
24	1.84E-06	27.0679	122.0977	-38.0549	32.09772	65.12281	-10.987	79.67635	29.27331
25	-4.26E-07	-42.8061	161.0264	-56.2841	71.02641	13.47797	-99.0902	79.78683	29.4418
26	5.10E-07	0.804883	119.2044	-91.7348	29.20436	92.53966	-90.9299	79.8613	29.56377
27	2.93E-07	9.469466	117.993	-77.5553	27.99301	87.02474	-68.0858	80.44904	29.06996
28	-4.64E-07	10.72079	108.4077	-67.2098	18.40767	77.93056	-56.489	80.2173	29.25883
29	7.92E-07	14.78235	120.7276	-68.7109	30.7276	83.49326	-53.9286	79.65318	29.73983
30	4.85E-07	68.3212	122.4197	-116.211	32.41966	184.5321	-47.8897	79.83462	29.96318
31	-1.16E-06	27.92151	109.0378	-119.019	19.03778	146.9404	-91.0974	79.91971	29.96864
32	7.26E-07	-11.1356	139.2819	-198.881	49.28187	187.7449	-210.016	78.94783	30.32082
33	6.86E-07	-22.1811	142.0982	-57.3622	52.09815	35.18111	-79.5433	78.6589	30.27227
34	9.12E-07	-40.4595	88.55891	-60.9931	178.5589	20.53359	-101.453	78.62523	30.38797
35	-1.32E-07	44.33593	112.9551	-91.363	22.95505	135.6989	-47.0271	79.89045	29.91392
36	-1.16E-06	19.0765	135.5199	-81.4348	45.51989	100.5113	-62.3583	78.75035	30.43268
37	1.42E-06	7.120198	84.47429	-134.444	174.4743	141.5642	-127.324	79.67702	30.6391
38	-7.74E-07	9.237582	122.8693	-159.482	32.86931	168.7197	-150.244	78.93326	30.62309
39	-1.59E-06	-2.95515	127.2037	-106.236	37.20367	103.2811	-109.191	78.77293	30.7093

S.NO.	Rotation ± uncertainty (degrees/yr)	Max horizontal extension (e1H) (nano-strain)	Azimuth of S1H (degrees)	Min horizontal extension (e2H) (nano-strain)	Azimuth of S2H (degrees)	Max shear strain (nano-strain)	Area strain (nano-strain)	Long (°)	Lat (°)
40	-6.95E-06	42.29392	23.21754	-174.236	113.2175	216.53	-131.942	78.7894	30.4542
41	-6.08E-06	195.2231	16.63236	-59.1451	106.6324	254.3682	136.078	78.85412	30.1137
42	-1.98E-06	119.6129	31.89535	-25.1974	121.8954	144.8103	94.41548	79.01866	29.94196
43	-5.04E-07	56.26886	150.589	-17.0123	60.58895	73.28111	39.25661	78.497	30.28397
44	3.95E-06	113.3468	126.2494	-103.518	36.24939	216.865	9.828563	78.44253	30.27166
45	1.83E-07	13.50632	159.8563	-8.96657	69.85626	22.47289	4.539758	78.26003	29.90215
46	2.17E-07	12.03511	154.3127	-12.1177	64.3127	24.21281	-0.1426	78.1004	29.99925
47	2.62E-06	47.60433	171.3537	-21.4498	81.35375	69.05415	26.1545	78.09562	30.2528
48	5.16E-07	-14.7983	116.051	-97.2158	26.051	82.41755	-112.014	78.40788	30.64831
49	-8.64E-07	5.991775	110.0822	-55.0814	20.0822	61.07316	-49.0896	79.02436	30.1249
50	-1.19E-07	29.3643	120.2667	-105.102	30.26673	134.4663	-75.7377	78.70181	30.65096
51	-9.85E-07	25.4069	126.6595	-60.1971	36.65949	85.60399	-34.7902	82.30015	28.3516
52	-1.36E-07	-2.76345	106.9935	-78.4504	16.99351	75.68695	-81.2138	82.72469	28.48492
53	-6.24E-08	-3.59438	104.1042	-81.4359	14.10417	77.84151	-85.0303	83.2421	28.20207
54	-6.48E-07	-7.67581	108.3235	-61.3681	18.32355	53.69227	-69.0439	83.30321	28.37135
55	-3.49E-07	-5.66473	108.4554	-74.8367	18.45539	69.17195	-80.5014	82.99606	28.43045
56	1.61E-09	-3.39801	104.8787	-82.0956	14.87865	78.6976	-85.4936	82.93539	28.261
57	5.42E-07	-7.34975	108.8526	-48.5599	18.8526	41.21012	-55.9096	83.02396	28.42762
58	3.15E-07	0.965328	112.7242	-49.3059	22.7242	50.27122	-48.3406	83.01617	28.58104
59	2.06E-07	-6.9486	116.768	-50.2089	26.76797	43.26029	-57.1575	83.295	28.37257
60	-9.76E-07	11.55383	130.9644	-62.7837	40.96443	74.33758	-51.2299	83.54244	28.38635

S.NO.	Rotation ± uncertainty (degrees/yr)	Max horizontal extension (e1H) (nano-strain)	Azimuth of S1H (degrees)	Min horizontal extension (e2H) (nano-strain)	Azimuth of S2H (degrees)	Max shear strain (nano-strain)	Area strain (nano-strain)	Long (°)	Lat (°)
61	-7.73E-07	-5.95005	136.1987	-64.1231	46.19874	58.17304	-70.0731	83.52032	28.34257
62	5.14E-07	-11.7672	110.0825	-48.6664	20.08247	36.89918	-60.4335	83.53281	28.25053
63	8.67E-07	-22.0907	98.77466	-48.6519	8.774663	26.56119	-70.7426	83.54998	28.20266
64	2.47E-07	-14.3185	100.9803	-54.7164	10.98025	40.39787	-69.0349	83.83905	28.19027
65	4.03E-07	-21.2864	87.3828	-68.9126	177.3828	47.62626	-90.199	83.85607	28.14237
66	6.75E-07	-6.42711	100.1659	-61.7546	10.1659	55.32747	-68.1817	83.76443	28.0608
67	6.77E-07	-9.64322	97.05148	-64.5422	7.051483	54.89894	-74.1854	83.54023	28.14369
68	-1.16E-07	-12.0751	113.4563	-44.9703	23.4563	32.89521	-57.0454	83.60176	28.31282
69	8.06E-07	-9.65619	99.36524	-67.3229	9.365238	57.66676	-76.9791	83.5472 9	27.99028
70	-1.41E-06	1.666171	111.5565	-227.039	21.55649	228.7048	-225.372	84.71106	27.99063
71	-1.77E-06	1.997162	113.5982	-185.794	23.59824	187.7912	-183.797	84.4879	28.075
72	1.82E-07	2.16203	107.8961	-121.065	17.89611	123.3265	-118.902	84.67058	27.75272
73	1.18E-06	0.762812	115.1201	-242.47	25.12014	243.2332	-241.708	84.89394	27.72048
74	4.45E-07	-19.6028	104.9755	-121.595	14.97547	101.9927	-141.198	84.74588	27.88156
75	1.12E-06	-47.9061	94.27342	-126.353	4.273422	78.44712	-174.259	84.76245	27.83363
76	-3.05E-06	22.5541	117.7994	-232.354	27.79936	254.9081	-209.8	84.97985	27.93243
77	-1.75E-06	-14.9844	127.9858	-273.508	37.98576	258.5236	-288.492	84.95728	27.88894
78	-5.29E-07	9.224338	109.923	-139.637	19.923	148.8609	-130.412	85.27155	27.75039
79	-4.47E-07	21.88203	104.3644	-123.731	14.36442	145.6126	-101.849	85.44619	27.80415
80	1.53E-06	-7.35197	97.29582	-308.868	7.295815	301.5161	-316.22	85.07896	27.83218
81	4.42E-07	-47.7993	90.07481	-333.606	0.074808	285.8065	-381.405	85.00853	27.84473

S.NO.	Rotation ± uncertainty (degrees/yr)	Max horizontal extension (e1H) (nano-strain)	Azimuth of S1H (degrees)	Min horizontal extension (e2H) (nano-strain)	Azimuth of S2H (degrees)	Max shear strain (nano-strain)	Area strain (nano-strain)	Long (°)	Lat (°)
82	4.87E-07	6.189758	104.0459	-253.488	14.04587	259.6779	-247.298	85.29538	27.84613
83	3.13E-07	10.3839	105.5967	-244.154	15.59665	254.5374	-233.77	85.41778	27.72976
84	1.55E-06	30.97821	99.49511	-228.562	9.495111	259.5402	-197.584	85.69234	27.81402
85	7.23E-07	7.642276	102.2797	-225.582	12.27966	233.2245	-217.94	85.51755	27.76057
86	2.64E-06	40.61847	108.5187	-236.438	18.51867	277.0563	-195.819	85.49479	27.71716
87	-2.53E-06	48.25904	100.9263	-128.556	10.92632	176.8148	-80.2968	86.44012	27.33266
88	3.76E-07	60.1606	92.38606	-135.83	2.386064	195.9902	-75.669	86.48312	27.59231
89	-1.19E-06	10.99028	100.9509	-113.997	10.95087	124.9877	-103.007	86.34737	27.25283

Table A1.
The estimated maximum & minimum principal strain rate, maximum shear strain rate, rotation rate, dilatation with their corresponding azimuth and coordinated by Triangulation approach.

Author details

Tandrila Sarkar[1], Abhishek Kumar Yadav[2], Suresh Kannaujiya[3*], Paresh N.S. Roy[2] and Charan Chaganti[3]

1 Indian Institute of Technology (Indian School of Mines), Dhanbad, Jharkhand, India

2 Indian Institute of Technology Kharagpur, West Bengal, India

3 Indian Institute of Remote Sensing, Indian Space Research Organization, Dehradun, Uttarakhand, India

*Address all correspondence to: skannaujiya@iirs.gov.in

IntechOpen

References

[1] Jacquelyne KW, Tilling RI. "The Story of Plate Tectonics." this Dynamic Earth. United States Geological Survey; 2016. DOI: 10.3133/7000097

[2] Lovett RA. Texas and Antarctica We re Attached, Rocks Hint. National Geographic News, Magazine; 2011

[3] Seeber L, Armbruster J. Great detachment earthquakes along the Himalayan arc and long-term forecasts. In: Simpson DW, Richards PG, editors. Earthquake Prediction: An International Review. Vol. 4. Washington DC: Am. Geophys. Union) Marice Ewing Series; 1981. pp. 259-277

[4] Bilham R, England P. Plateau 'pop-up' in the great 1897 Assam earthquake. Nature. 2001;**410**(6830):807-809. DOI: 10.1038/35071057

[5] Kannaujiya S, Yadav RK, Champati ray PK, Sarkar T, Sharma G, Chauhan P. Unraveling seismic hazard by estimating prolonged crustal strain buildup in Kumaun-Garhwal, northwest Himalaya using GPS measurements. Journal of Asian Earth Sciences. 2022;**223**:104993. DOI: 10.1016/j.jseaes.2021.104993

[6] Yadav RK, Gahalaut VK, Bansal AK, Sati SP, Catherine J, Gautam P, et al. Strong seismic coupling underneath Garhwal–Kumaun region, NW Himalaya, India. Earth and Planetary Science Letters. 2019;**506**:8-14. DOI: 10.1016/j.epsl.2018.10.023

[7] Rajendran CP, John B, Anandasabari K, Sanwal J, Rajendran K, Kumar P, et al. On the paleoseismic evidence of the 1803 earthquake rupture (or lack of it) along the frontal thrust of the Kumaun Himalaya. Tectonophysics. 2018;**722**:227-234

[8] Banerjee P, Bürgmann R, Nagarajan B, Apel E. Intraplate deformation of the Indian subcontinent.

Geophysical Research Letters. 2008;**35**(18). DOI: 10.1029/2008GL035468

[9] Le Fort P. Himalaya: The collided range. Present knowledge of the continental arc. American Journal of Science. 1975;**275A**:1-44

[10] Caldwell WB, Klemperer SL, Lawrence JF, Rai SS, Ashish. Characterizing the Main Himalayan thrust in the Garhwal Himalaya, India with receiver function CCP stacking. Earth and Planetary Science Letters. 2013;**367**:15-27

[11] Rawat G, Arora BR, Gupta PK. Electrical resistivity cross-section across the Garhwal Himalaya: Proxy to fluid-seismicity linkage. Tectonophysics. 2014;**637**:68-79

[12] Gilligan A, Priestley KF, Roecker SW, Levin V, Rai SS. The crustal structure of the Western Himalayas and Tibet. Journal of Geophysical Research - Solid Earth. 2014;**120**(5):3946-3964

[13] Lyon-Caen H, Molnar P. Gravity anomalies, flexure of the Indian plate, and the structure, support and evolution of the Himalaya and Ganga Basin. Tectonics. 1985;**4**(6):513-538

[14] Lang KA, Huntington KW, Burmester R, Housen B. Rapid exhumation of the eastern Himalayan syntax is since the late Miocene. Geological Society of America Bulletin. 2016;**128**(9–10):1403-1422

[15] Sorkhabi RB, Macfarlane A. Himalaya and Tibet: Mountain roots to mountain tops. Geological Society of America Special Paper. 1999;**328**:1-7

[16] Kannaujiya S, Philip G, Kannaujiya S, Yadav RK, Champati ray PK, Sarkar T, Sharma G, Chauhan P. Integrated geophysical techniques for subsurface imaging of active deformation across

the Himalayan frontal thrust in Singhauli, kala Amb, India. Quaternary International. 2021;**575–576**:72-84. DOI: 10.1016/j.quaint.2020.05.003

[17] Yadav RK, Gahalaut VK, Bansal AK. Tectonic and non-tectonic crustal deformation in Kumaun Garhwal Himalaya. Quaternary International. 2021;**585**:171-182. DOI: 10.1016/j. quaint.2020.10.011

[18] Sharma G, Kannaujiya S, Gautam PKR, Taloor AK, Champatiray PK, Mohanty S. Crustal deformation analysis across Garhwal Himalaya: Part of western Himalaya using GPS observations. Quaternary International. 2021;**575**:153-159. DOI: 10.1046/j.1365-246X.2003.01917.x

[19] Jade S, Shrungeshwara TS, Kumar K, Choudhury P, Dumka RK, Bhu H. India plate angular velocity and contemporary deformation rates from continuous GPS measurements from 1996 to 2015. Scientific Reports. 2017; 7(1). DOI: 10.1038/s41598-017-11697-w

[20] Herring TA, King RW, McClusky SC. Introduction to GAMIT/ GLOBK, Release 10.4. Cambridge, MA: Dept. of Earth Atmos. and Planet. Sci., Mass. Instit. of Technol.; 2010. p. 48

[21] Boehm J, Heinkelmann R, Schuh H. Short note: A global model of pressure and temperature for geodetic applications. Journal of Geodesy. 2007; **81**:679-683

[22] Frank F. Deduction of earth strains from survey data. Bulletin of the Seismological Society of America. 1966; **56**(1):35-42

[23] Bibby HM. Crustal strain from triangulation in Marlborough. New Zealand. Tectonophysics. 1975;**29**: 529–540

[24] Cronin VS, Resor PG, Hammond WC, Kreemer C, Olds SE,

Pratt-Sitaula B, et al. Developing a curricular module for introductory geophysics or structural geology courses to quantify crustal strain using earth scope PBO GPS velocities. In: Abstract ED41B-0681, Fall Meeting. San Francisco: AGU; 2012

[25] Sreejith KM, Sunil PS, Agrawal R, Saji AP, Rajawat AS, Ramesh DS. Audit of stored strain energy and extent of future earthquake rupture in central Himalaya. Scientific Reports. 2018;**8**(1). DOI: 10.1038/s41598-018-35025-y

[26] Kreemer C, Holt WE, Haines AJ. An integrated global model of present-day plate motions and plate boundary deformation. Geophysical Journal International. 2003;**154**(1): 8-34. DOI: 10.1046/j.1365-246x. 2003.01917.x

[27] Joshi GR, Hayashi D. Development of extensional stresses in the compressional setting of the Himalayan thrust wedge: Inference from numerical modelling. Natural Science. 2010; **02**(07):667-680. DOI: 10.4236/ ns.2010.27083

[28] Norton M. Late Caladonian extension in western Norway: A response to extreme crustal thickening. Tectonics. 1986;**5**(2):192-204

[29] Corredor F. Eastward extend of the late eocene early oligocene onset of deformation across the northern Andes: Constraints from the northern portion of the eastern cordillera fold belt, Colombia. Journal of South American Earth Sciences. 2006;**16**(6):445-457

[30] Royden LH, Burchfiel BC. Thin skinned N-S extension within the convergence Himalayan region; gravitational collapse of a Miocene topographic front. In: Croward MP, Dewey JF, Hanback PL, eds. Continental Extensional Tectonics. Vol. 26(5-6) London: Geological Society; 1987. pp. 611-619

[31] Lave J, Avouac JP. Active folding of fluvial terraces across the Siwalik Hills, Himalayas of Central Nepal. Journal of Geophysical Research. 2000;105(B3): 5735-5770

[32] Joshi GR, Hayashi D. Finite element modelling of the pull-apart formation: Implication for tectonics of Bengo Co pull-apart basin, Southern Tibet. Natural Science. 2010;2(6):654-666

[33] Thakur VC. Active tectonics of Himalayan frontal fault system. International Journal of Earth Sciences. 2013;102(7):1791-1810

[34] Savage JC. Strain accumulation in Western United States. Annual Review of Earth and Planetary Sciences. 1983;11: 11-43

[35] Okada Y. Surface deformation due to shear and tensile faults in a half-space. Bulletin of the Seismological Society of America. 1985;75:1135-1154

[36] Mao A, Christopher GA, Timothy HD. Noise in GPS coordinate time series. Journal of Geophysical Research. 1999;104:2792-2816

[37] Kumahara Y, Jayangondaperumal R. Paleoseismic evidence of a surface rupture along the Northwestern Himalayan frontal thrust (HFT). Geomorphology. 2013;180(181):47-56

[38] Yadav A, Kannaujiya S, Champati Ray PK, Yadav RK, Gautam PK. Estimation of crustal deformation parameters and strain build-up in northwest Himalaya using GNSS data measurements. Contributions to Geophysics and Geodesy. 2021;51(3): 225-243. DOI: 10.31577/congeo.2021. 51.3.2

[39] Lindsey EO, Almeida R, Mallick R, Hubbard J, Bradley K, Tsang LLH, et al. Structural control on Downdip locking extent of the Himalayan megathrust. Journal of Geophysical Research - Solid

Earth. 2018;123(6):5265-5278. DOI: 10.1029/2018JB015868

[40] Dal Zilio L, Hetényi G, Hubbard J, Bollinger L. Building the Himalaya from tectonic to earthquake scales. Nature Reviews Earth & Environment. 2021; 2(4):251-268. DOI: 10.1038/s43017-021-00143-1

[41] Elliott J et al. Himalayan megathrust geometry and relation to topography revealed by the Gorkha earthquake. Nature Geoscience. 2016;9:174

[42] Bilham R. Earthquakes in India and the Himalaya: Tectonics, geodesy and history. Annales de Geophysique. 2004; 47:2-3

[43] Stevens V, Avouac JP. Millenary mw > 9.0 earthquakes required by geodetic strain in the Himalaya. Geophysical Research Letters. 2016;43:1118-1123

[44] Armijo R, Tapponnier P, Mercier J, Han TL. Quaternary extension in southern Tibet: Field observations and tectonic implications. Journal of Geophysical Research - Solid Earth. 1986;91:13803-13872

[45] Dal Zilio L, van Dinther Y, Gerya T, Avouac JP. Bimodal seismicity in the Himalaya controlled by fault friction and geometry. Nature Communications. 2019;10(1). DOI: 10.1038/s41467-018-07874-8

[46] Rizza M et al. Post earthquake aggradation processes to hide surface ruptures in thrust systems: The M8.3, 1934, Bihar-Nepal earthquake ruptures at Charnath Khola (eastern Nepal). Journal of Geophysical Research - Solid Earth. 2019;124:9182-9207

[47] Rajendran CP, John B, Rajendran K. Medieval pulse of great earthquakes in the central Himalaya: Viewing past activities on the frontal thrust. Journal of Geophysical Research - Solid Earth. 2015;120:1623-1641

Seismic Forecasting Using a Brownian Passage Time Distribution

Edmore Utete

Abstract

Seismic forecasting using a Brownian Passage Time distribution is presented in this chapter. Seismic forecasting is concerned with the probabilistic assessment of general seismic hazard, including the frequency and magnitude of earthquakes in a given area over a given period of time. Seismic forecasting generally look for trends that lead to an earthquake. The estimation of the time that a strong earthquake will occur requires the determination of the distribution that the earthquake recurrence time follows. Brownian Passage Time distribution describes reliably the physical processes related with earthquakes' occurrence. The model assumes that the evolution of the stress loading between two earthquakes depends on the constant loading rate, and a random component, which follows the Brownian Relaxation Oscillator. Its hazard function is in good agreement with the temporal evolution of earthquake occurrence as the hazard rate is very low after an earthquake, and then increases as time passes, and takes a maximum value at the mean recurrence time and since then, it decreases asymptotically exhibiting a pure quasi-periodic temporal occurrence.

Keywords: seismology, seismic forecasting, probability distribution, stochastic model

1. Introduction to seismic forecasting

Seismic forecasting is concerned with the probabilistic estimation of the frequency and magnitude of seismic events in a given area over a given period of time using advanced statistical and scientific methods. This can be distinguished from seismic prediction, which is the specification of the time, location, and magnitude of a future seismic event with sufficient precision that a warning can be issued. The two can be further distinguished from seismic warning systems, which is the detection of a possible seismic event in real time to regions that might be affected. Seismic warning systems focus on a very short time outlook ranging from few minutes to few days, seismic prediction looks at specific future time with acceptable time ranging from few minutes to few hours while seismic forecasting makes use of probability estimations with time scales ranging from few days up to several decades.

In this chapter on seismic forecasting, literature body on seismic formation process will be reviewed to develop a theoretical framework that validates and upholds ideas that will be further used to develop a stochastic model that can be used to forecast future seismic events. The elastic rebound theory will be validated, and an analysis of recurrence time models will be done so as to select the best model that can be used in seismic forecasting. Given a homogeneous, consistent, and complete past seismic data of a region, the unbiased maximum-likelihood estimates of model parameters can be estimated and used as input parameter to seismic forecasting.

1.1 History of seismic forecasting

Since the ancient times attempts to predict seismic events were made, with people associating such events with the spiritual world. In some societies, such events were considered a sign of bad luck or punishment for disobedience from the supernatural beings. The scientific revolution was a game changer in this mistrial subject with scientist being optimistic that a practical method of seismic prediction would soon be found. By the end of the nineteenth century continued failure leads to many people questioning whether it was even possible to predict a seismic event. Scientific evidence of few predictions of large seismic events has not occurred, and a few claims of success prediction remain controversial. As a result, emphasis has been shifted from seismic prediction to seismic forecasting. Due to the high level of destruction and loss of lives, after larger seismic events, a lot of scientific and national government resources have been pooled and allocated into seismic forecasting rather than prediction as it has proved to be useful in seismic risk mitigation in areas such as establishment of building codes, insurance rate structures, awareness, preparedness programs, and public policy related to seismic events. Statistical methods used for seismic forecasting look for trends or patterns that lead to a seismic event. The trends involve many complex variables, and the advanced statistical techniques are needed to understand them. These approaches tend to have relatively long time periods, making them useful for seismic forecasting.

1.2 Elastic-rebound model

Previously, it was thought that ruptures of the surface were the result of strong ground shaking rather than the converse suggested by Harry Fielding Reid [1], the first scientist to explain the seismic formation process after the great 1906 San Francisco earthquake. The theory postulated that steady tectonic force causes strain to accumulate slowly in a rock and eventually become large up to a threshold constant value called the elastic limit. The elastic limit is the maximum strain that a rock can withstand without breaking. When the limit is exceeded, an earthquake will occur. At that time, a sudden movement occurs along the fault line, releasing the accumulated energy, and the rocks snap back to their original undeformed shape. After an event another cycle starts. When the accumulated strain is great enough to overcome the strength of the rocks, an earthquake occurs again. The duration of an "earthquake cycle" is the ratio of event-strain release to tectonic strain rate. Rocks in a fault plane are subjected to tectonic force caused by tectonic plate movement. Fault plane subjected to plate tectonic moves at the rate of a few centimeters per year, over a time period of decades. The stored energy is released during the rupture partly as heat, partly in damaging the rock, and partly as seismic waves.

The elastic-rebound model plays a central acceptable role in our understanding of earthquake mechanics. This approach, which has some observational grounding, has been the basis for long-term forecasting models. A number of statistical models have been proposed for seismic forecasting. Discrete probability models are proposed to forecast a number of events in a given time interval and continuous probability models have been proposed to forecast the time until the next seismic event.

It is not a good idea to model seismic recurrence using a normal distribution as it gives positive probability to negative intervals. Better models are the Weibull gamma and log-normal distributions, and can be used as alternatives. In this chapter, we are going to consider the Weibull distribution as it is practically convenience as proven by its wide application in statistical quality control. Nishenko and Buland [2] identified several theoretical distributions after normalizing 15 characteristic seismic sequences data by looking at the shape of a generic distribution for recurrence intervals.

Nishenko and Buland [2], agreed with Hagiwara's [3] model preferences of the exponential, Weibull, gamma, and lognormal distributions by analyzing the reliability of the distributions. They further pointed that the log-normal provides the best fit to the distribution of normalized intervals, and there was no difference in the estimated parameters of the log-normal normal distribution in seismic data from different regions with different time scales. They all postulated that log-normal distribution was a good model for seismic risk assessment.

The elastic rebound model provides a framework seismic recurrence modeling. Extending advanced probabilistic modeling to this theory provides a basic model for seismic forecasting. The idea is the inclusion of the seismic random perturbations in the elastic rebound model.

The same approach was used by Kagan and Knopoff [4] in modeling time-dependent model in statistical seismic analysis. Empirical analysis does not use a Poisson process in many cases. Matthews et al. [5] disagreed with Kagan and Jackson [6] on the idea that seismic recurrence time intervals are shorter than the mean interval. The use of exponential distribution in modeling seismic recurrence is debatable, and currently, there is no much literature and it is still not clear whether seismic recurrence has a strong central tendency or not.

1.3 Recurrence time models

The recurrence time models that can be considered are the exponential, Weibull, gamma, and log-normal distributions. Use of these distributions has been motivated primarily on the grounds of familiarity, simplicity, and convenience.

1.3.1 Exponential distribution

The exponential distribution is the probability distribution of the time between seismic events in a Poisson point process. A number of events in a given time interval follow a poison distribution as it assume that events occur continuously and independently at a constant average rate. This means the time between events follows an exponential distribution. One-parameter exponential distribution has a property of memoryless implying the distribution of a waiting time until a certain event does not depend on how much time has elapsed already. This property disqualifies the exponential distribution as a possible model for seismic recurrence time model as it opposes the elastic rebound model.

1.3.2 Gamma distribution

Gamma distribution is a generalized exponential distribution. This distribution contains the densities of the sum of n independent exponentials. The shape parameter $\theta > 1$ has zero hazard rate at a time zero and increases to a finite asymptotic level that is always smaller than the mean recurrence rate, which is not the case with seismic events. This makes the Gamma distribution not a potential model for seismic recurrence time models.

1.3.3 Weibull distribution

Weibull distribution contains the densities of the minimum of n independent exponential distribution with rate t that have independent occurrence times, and then, the distribution has a Weibull (t, n). It is a distribution for which seismic occurrence rate is proportional to a power of time. The shape parameter, k, is that power plus one. A value of $k < 1$ indicates that the seismic occurrence rate decreases over time that is there is significant "infant mortality," or seismic event occurrence is high early and decreasing over time as the seismic events are "weeded" out of the population. This may be because the seismic building force stabilizes over time. A value of $k = 1$ indicates that the seismic occurrence rate is constant over time. This is consistent with the elastic rebound model but the Weibull distribution reduces to an exponential distribution. A value of $k > 1$ indicates that the seismic occurrence rate increases with time. This happens if there is an increase in seismic building process activities. The hazard-rate functions either start at zero and increase to ∞ or *vice versa*, depending on the parameter k, this disqualifies the weibull distribution as a potential seismic recurrence time model as the elastic rebound model do not support infinite or zero hazard rate.

1.3.4 Log-normal distribution

The log-normal distribution is obtained by taking the exponential of a normal distribution. The asymptotic hazard rate is always zero and hazard-rate functions that increase from zero at time $t = 0$ and then eventually decrease to zero. The probability density function puts least weight in the left tail. The mean residual life increases without bound as t → ∞. Its asymptotic properties disqualify the log-normal family as a reliable seismic recurrence model as this means the longer the time since the last seismic event, the longer it will be expected until the next seismic event. As suggested by Davis et al. [7] that the log-normal model provided only a slightly better fit than the gamma or Weibull models.

1.3.5 The Brownian relaxation oscillator

The Brownian Relaxation Oscillator can be used for seismic forecasting of time of next event. If seismic building process are fixed and tectonic forces load at a constant rate, seismic events will occur after a fixed time interval. The point process will form identical events. The only significant variable in such a deterministic model will be the "strain state," and after a long time of events, it forms a cycle of loading and instant relaxation oscillating over time forming a deterministic relaxation oscillator model. The strain state will go to zero soon after a seismic event and increase upward at a constant rate up to a fixed elastic limit value. Immediately after exceeding that value an event occurs that relaxes the strain level to zero. The cycle will continue over time making the time of next seismic event predictable.

This is in agreement with the elastic rebound model proposed by Reid [1] with strain level meaning the same as cumulative elastic strain. The strain level could also mean cumulative moment deficit or total stress level. The strain level can be described as the absolute rapture potential. **Figure 1** shows a diagrammatic representation of a deterministic relaxation oscillator.

Let $S(t)$ be the strain in the fault plane at time t measured from a value S_0 after an event. The strain in the fault is the sum of initial strain after an event S_0 and an increase in strain φ due to tectonic loading. The elastic rebound model assumes the strain in the rock load constantly at a rate λ. An event will occur when stain level reaches a constant elastic limits value S_E.

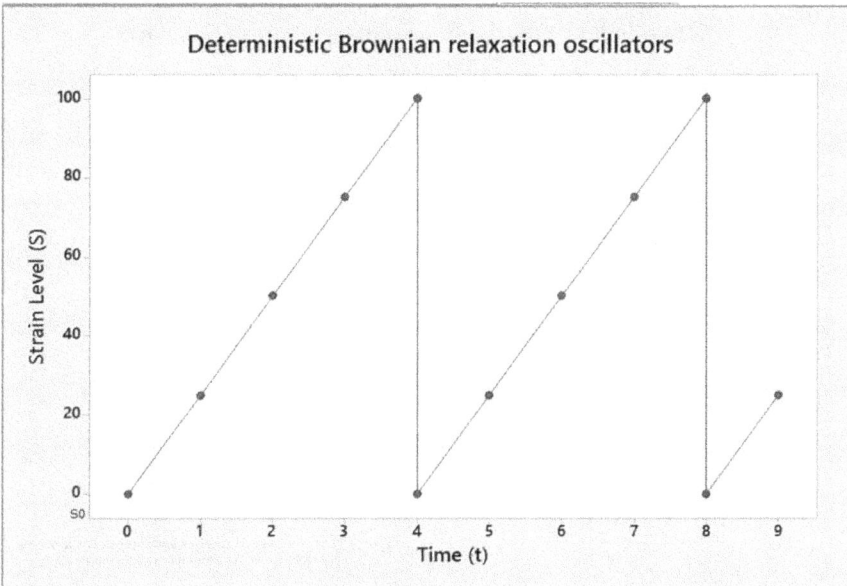

Figure 1.
Deterministic relaxation oscillator.

$$\varphi = \lambda t \tag{1}$$

$$S(t) = s_0 + \lambda t \tag{2}$$

Since φ will be set to 0 after an event and starts to increase constantly at a rate λ until the next event. If events occur at constant intervals t_E and if t is the clock time, then

$$S(t) = s_0 + \lambda(t - t_E) \tag{3}$$

$$S(t) = s_0 + \lambda t - \lambda t_E \tag{4}$$

The deterministic process can be expressed as

$$S(t) = s_0 + \varphi t - \varphi t_E \tag{5}$$

The graphical representation of such a process is shown in **Figure 1** aforesaid.

The seismic formation process has other complex variables, which can be represented in the model with a random error term or white noise εt a random perturbation process. The stochastic relaxation oscillator can be expressed as

$$S(t) = s_0 + \varphi t - \varphi t_E + \varepsilon_t \tag{6}$$

The above stochastic relaxation oscillator equations show that the stain level is made up of a sum of three components.

s_o is the initial constant stain pre-existing in a fault plane. The value of s_o is a constant but differs from fault plane to faults depending on the type of rock.

$\varphi t - \varphi t_E$ is the difference between the strain level at any given time and the strain level after the previous seismic event. The value is as a result of stain accumulation due to seismic building process and depends on time since the last seismic event. Seismic forecasting involves being able to find the best probability model that best fits this component. The component is assumed to follow a Brownian

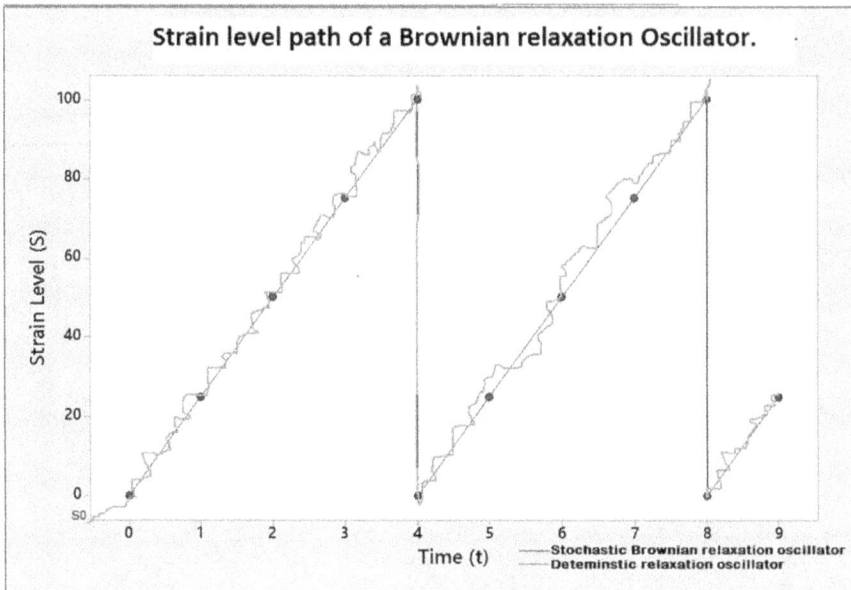

Figure 2.
Brownian relaxation oscillator with random term.

Passage-Time Distributions as it depends on time past since the last event. In this model $s_0 \leq \varphi t_E$, this explains why some events may result in aftershock or event after of aftershock because in some events the relaxation may not release all the stains in the fault giving a possibility of an event after another event.

The component ε_t is a random term that takes into the model the effect of other random variables that affect seismic stain building as a result of seismic perturbation. The term is also called the white noise and follows a Brownian motion or Gaussian or normal distribution with mean zero and constant variance. This term results in the formation of a graph with a path up or above the smooth theoretical graph of S_t.

The graphical representation of such a process is shown in red color in **Figure 2**. Such a path is like the drunken man's path. The value of constant variance determines how far the process moves from the line representing the deterministic model.

The seismic strain builds between the time interval t_0 and t_E and builds at a rate. This means the seismic recurrence intervals will have average length $\mu = \frac{\delta}{\lambda}$. Indeed, the deterministic oscillator with identical recurrence intervals has a variance, $\sigma = 0$.

1.4 Brownian passage-time distributions

Let T denote the first passage time to level $s > 0$ by Brownian motion with drift rate $\lambda > 0$ and diffusion rate σ^2. The probability distribution of T has a well-known closed form with probability density function given by

$$f(t; s; \lambda; \sigma) = \frac{s}{\sqrt{2\pi\sigma^2 t^3}} \exp\left(-\frac{(s - \lambda t)^2}{2\sigma^2 t}\right), t \geq 0 \qquad (7)$$

The distribution is also called the inverse Gaussian or the Brownian passage time.

The cumulative Gaussian probability function is given by

$$F(t) = P(T \leq t) \tag{8}$$

Let $\alpha = \dfrac{\sigma}{\mu}$

$$= \Phi\left\{ \frac{1}{\alpha}\left(\sqrt{\frac{t}{\mu}} - \sqrt{\frac{\mu}{t}} \right) \right\} + exp^{\left(\frac{2}{\alpha^2}\right)} \Phi\left\{ \frac{1}{\alpha}\left(\sqrt{\frac{t}{\mu}} + \sqrt{\frac{\mu}{t}} \right) \right\} \tag{9}$$

Using the log-likelihood function is given by

$$L(s; \lambda; \sigma/t_i) = \prod_{i=1}^{n}\left\{ \frac{s}{\sqrt{2\pi\sigma^2 t_i{}^3}}\, \exp\left(-\frac{(s - \lambda t_i)^2}{2\sigma^2 t_i} \right) \right\} \tag{10}$$

we find that expectation of T, $E(T) = \mu$ and variance of T, $var(T) = \sigma^2 = (\mu\alpha)^2$ where α is the ratio of the standard deviation to the mean, that is, the coefficient of variation or the aperiodicity of the failure-time distribution. This is also called the ratio of the sample standard deviation to the sample mean. α is a measure of irregularity in the seismic event sequence and determines the shape of the Brownian Passage-Time Distributions.

1.4.1 Time dependence

According to Matthews et al. [5], the Brownian passage-time distributions quantify occurrence-time probabilities for a steadily loaded system subject to random perturbations. The distribution is used to answer questions like "what is the effect of past time since the last event on conditional probabilities for the next event?"

1.4.2 Hazard rate

The hazard rate of Brownian passage-time distributions is given by

$$h(t) = \frac{f(t)}{1 - F(t)} \tag{11}$$

1.4.3 Residual life distribution

To examine the effect of elapsed time on occurrence probabilities, we may consider the residual life conditioned on $> t$. The conditional probability of a seismic event at any given time, t given that time P has passed after an event is given by

$$R(P) = P(P \leq t + P/P > t) \tag{12}$$

$$= \frac{F(t + P) - F(t)}{1 - F(t)} \tag{13}$$

residual life density

$$= r(p) = \frac{\partial R(P)}{\partial P} \tag{14}$$

$$= \frac{f(t + P)}{1 - F(t)} \tag{15}$$

mean residual life is

$$m(p) = \int_0^\infty p \, r(p) \partial p \tag{16}$$

The properties of the Brownian passage-time distributions that makes it appropriate for seismic forecasting is the shape of the hazard-rate function and behavior of the residual life as time increases, asymptotic mean residual life, and hazard rate. The value of $h(0)$ is also important, as it governs the likelihood of immediate re-rupture after an event. All BPT hazard-rate functions share a general shape, The hazard-rate function, $h(t)$, of all models in the Brownian passage-time distribution family always starts at 0 at $t = 0$, then goes upward a highest value at some time after the probability distribution's mode, and then goes down approaching a asymptotically a fixed value h_∞ as time approaches infinity. The value can be found by taking the limit as $t \to \infty$ as,

$$h_\infty = \frac{1}{2\mu\alpha^2} \tag{17}$$

Brownian failure process eventually attains a quasi-stationary state in which residual time to failure becomes independent of passed time an applicable property in seismic forecasting. The Brownian passage-time distribution describes the failure probability of a Brownian relaxation oscillator as a function of elapsed time and the statistical properties of the failure time series.

Brownian relaxation oscillator can be used to estimate the probability of a seismic event in the next time interval in a given region. Seismic data from the past events can be utilized to estimate the parameters $s; \lambda; \sigma$, which differs from region to region. Once the parameter has been estimated a good approximation of t, the time until the next event can then be calculated. This is the proposed model for seismic forecasting.

1.5 Model parameter estimation

1.5.1 Maximum-likelihood estimates

Suppose we have a consistent, complete, and homogeneous past seismic data of time between seismic events, $t_1; t_2; t_3; \ldots .t_n$ of a given region. The estimates of model parameters $s; \lambda$ and σ can be found by looking for values that maximize the parameters using the sample data. Such values can be found by equating to zero the first derivatives of $f(t;s;\lambda;\sigma)$ with respect to each parameter.

$$\frac{\partial f(t;s;\lambda;\sigma)}{\partial s} = \prod_{i=1}^{n} \frac{\partial \frac{s}{\sqrt{2\pi\sigma^2 t_i{}^3}} \exp\left(-\frac{(s-\lambda t_i)^2}{2\sigma^2 t_i}\right)}{\partial s} = 0 \tag{18}$$

$$\frac{\partial f(t;s;\lambda;\sigma)}{\partial \lambda} = \prod_{i=1}^{n} \frac{\partial \frac{s}{\sqrt{2\pi\sigma^2 t_i{}^3}} \exp\left(-\frac{(s-\lambda t_i)^2}{2\sigma^2 t_i}\right)}{\partial \lambda} = 0 \tag{19}$$

$$\frac{\partial f(t;s;\lambda;\sigma)}{\partial \sigma} = \prod_{i=1}^{n} \frac{\partial \frac{s}{\sqrt{2\pi\sigma^2 t_i{}^3}} \exp\left(-\frac{(s-\lambda t_i)^2}{2\sigma^2 t_i}\right)}{\partial \sigma} = 0 \tag{20}$$

The aforesaid equations will give maximum-likelihood estimates of $s; \lambda$ and σ that can be used in the probability density function $f(t; s; \lambda; \sigma)$ to estimate the variable time t of future events.

1.5.2 Magnitude forecast

The time of a seismic event is estimated from the above model $f(t; s; \lambda; \sigma)$, but every seismic event must have a forecasted value of the expected magnitude. Suppose we have consistent, complete and homogeneous past seismic data of magnitudes $m_1; m_2; m_3; \ldots .. m_n$ of a given region.

The unbiased estimate of the future expected magnitude is the mean of past magnitude

$$m_\mu = \frac{\sum_{i=1}^n m_i}{n} \tag{21}$$

From 1.5.1 and 1.5.2 above, the time and magnitude of future events can be forecasted.

2. Conclusion

The Brownian relaxation oscillator with random term for seismic random perturbation modeled above represents a model that can be used for seismic forecasting. The inclusion of the error term in the model gives allowance of a deviation of the forecast from the actual event. Since the error term is known to be normally distributed with mean zero and constant variance, the expectation of the deviation is zero. The error term also represents other seismic formation variables and errors in model parameter estimation.

For one to accurately use the model, a consistent, complete, and homogeneous with unified magnitude, past seismic data are needed. Such data are always unavailable, because a complete catalog of the full population of events begins from the start of the earth planet formation. Statistical method utilizes a very small sample beginning "yesterday" when seismic monitoring started to infer population parameters. If such model parameters can be accurately estimated, seismic events can be accurately forecasted using the Brownian relaxation oscillator with random perturbation.

Conflict of Interest

The author declares no conflict of interest.

Author details

Edmore Utete
National University of Science and Technology, Bulawayo, Zimbabwe

*Address all correspondence to: eddieutete@yahoo.com

IntechOpen

References

[1] Reid HF. On mass-movements in tectonic earthquakes. In: The California Earthquake of April 18, 1906: Report of the State Earthquake Investigation Commission. Washington, D.C: Carnegie Institution of Washington; 1910

[2] Nishenko S, Buland R. A generic recurrence interval distribution for earthquake forecasting. Bulletin of the Seismological Society of America. 1987; 77(1382):1389

[3] Hagiwara Y. Probability of earthquake occurrence as obtained from a Weibull distribution analysis of crustal strain. Tectonophysics. 1974;23:313-318

[4] Kagan YY, Knopoff L. Statistical short-term earthquake prediction. Science. 1987;236(4808):1563-1567

[5] Matthews MV, Ellsworth WL, Reasenberg PA. A Brownian model for recurrent earthquakes. Bulletin of the Seismological Society of America. 2001; 92(6):2233-2250

[6] Kagan YY, Jackson DD. Worldwide doublets of large shallow earthquakes. Bulletin of the Seismological Society of America. 1999;89:1147-1155

[7] Davis PM, Jackson DD, Kagan YY. The longer it has been since the last earthquake the longer the expected time till the next? Bulletin of the Seismological Society of America. 1989; 79:1439-1456

Chapter 9

Revelation of Potentially Seismic Dangerous Tectonic Structures in a View of Modern Geodynamics of the Eastern Caucasus (Azerbaijan)

Talat Kangarli, Tahir Mammadli, Fuad Aliyev,
Rafig Safarov and Sabina Kazimova

Abstract

The stress state of the earth's crust in the Eastern Caucasus, located in the zone of collision junction of the North Caucasian, South Caucasian, and Central Iranian continental massifs, is a consequence of the inclusion of the Arabian indenter into the buffer structures of the southern framing of Eurasia at the continental stage of alpine tectogenesis. This evidenced from the results of geophysical observations of the structure and seismic-geodynamic activity of the region's crust. The latter, at the neotectonic stage, was presented as underthrust of the South Caucasian microplate under the southern structures of Eurasia. The analysis and correlation of historical and recent seismic events indicate the confinement of most earthquake foci to the nodes of intersection of active faults with various orientations or to the planes of deep tectonic ruptures and lateral displacements along unstable contacts of material complexes of various competencies. The focal mechanisms of seismic events reveal various rupture types, but in general, the earthquake foci are confined to the nodes of intersection of faults of the general Caucasian and anti-Caucasian directions. Based on the observed weak seismicity, active areas of deep faults were identified, which are accepted as potential source zones.

Keywords: earthquake, seismotectonics, focal mechanism, geodynamics, accretionary prism

1. Introduction

The territory of the Middle East, the northern periphery of which corresponds to the South Caucasus, is a collage of different-scale tectonic blocks—Anatolian-Taurus, Central Iranian, South Caucasian microplates, and smaller blocks (**Figure 1**), located between the Arabian continental plate (in the south) and the southern edge of the Eurasian continent (in the north). The latter at the neotectonic stage of tectogenesis (from the end of the Miocene) exist in the regime of collision convergence, which in turn causes exceptional tectonic activity in the region [2–15]. This feature is evidenced by often occurrence of strong and destructive earthquakes in Turkey, Iran, and the Caucasus Isthmus in the present time. The seismicity of these territories is explained by intensive restructuring of the structural plan with significant amplitudes of recent movements.

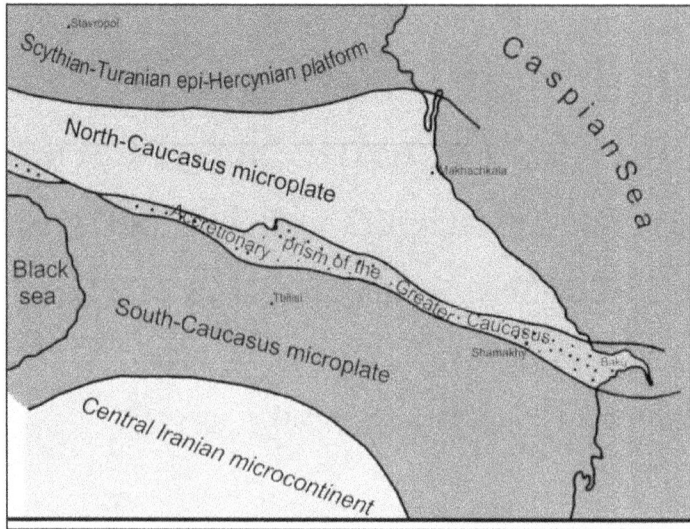

Figure 1.
Allocation of accretion prism within structure of the Greater Caucasus of the Caucasus isthmus (modified from [1]).

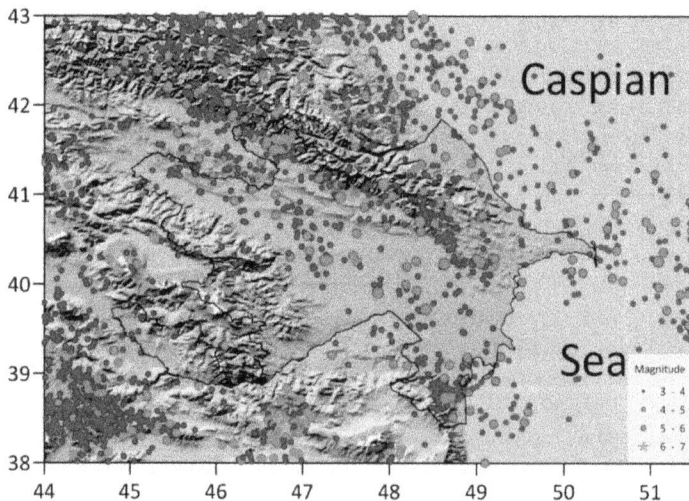

Figure 2.
Map of earthquakes epicenters $M \geq 3.0$ of the territory of Azerbaijan for the period 2004–2020.

In this regard, the eastern part of the South Caucasus, where Azerbaijan is located, characterized by highly seismic activity with periodic occurrence of seismic events with $M > 5$, is no exception (**Figure 2**).

The stress state of the earth's crust in the region located in the collision junction zone of the North Caucasian, South Caucasian, and Central Iranian continental massifs (tectonic microplates) is a consequence of the intrusion of the Arabian indenter into the buffer structures of the southern framing of Eurasia at the continental stage of alpine tectogenesis (from the end of the Miocene). This is evidenced by the results of geophysical observations of the structure and seismic-geodynamic activity of the region's earth crust. The latter, at the neotectonic stage, was an underthrust (S-subduction—continental subduction or pseudosubduction) region of the South Caucasian microplate under the southern underbelly of Eurasia (Scythian-Turanian epigercynian platform) in the northern wing and active terrestrial volcanism with the formation of volcano-plutonic complexes. Namely, the process of lateral compression, which continues at the current stage of tectogenesis under the influence of the collision approach of the Arabian and Eurasian continents, determines the high level of seismic and geodynamic activity in the study area.

Seismological and paleoseismotectonic studies, and seismic and seismotectonic zoning works carried out in various seismic regions of the Caucasus (including territory of Azerbaijan) confirm the controllability of earthquake focal areas by a network of faults of general Caucasian and anti-Caucasian direction with various types of prolongation. However, in general, the reason of current seismic activity is the horizontal movements of different-scale tectonic blocks of the earth's crust, located in the zone of collision interaction of the Afro-Arabian and Eurasian continental plates.

We carried out the analysis and interpretation of seismological data, as well as the results from GPS monitoring of modern geodynamic activity with the identification of their correlations with the features of the deep structure. GPS monitoring data in the Eastern Caucasus indicate an intensive advancement of the South Caucasus block in the northern points. The analysis and correlation of historical and recent (2012–2020) seismic events indicate the confinement of earthquake sources mainly to the nodes of intersection of active faults of various strikes or to the planes of deep tectonic disruptions and lateral displacements along unstable contacts of material complexes of various competencies.

2. Recent geodynamic processes

The observed seismic activity is generally confined with the rates of horizontal movements that took place for the period of GPS monitoring of the modern geodynamics of the region since 1998 [6, 16–21]. In comparison with the data for 2004, the rates of horizontal movements for the absolute majority of observation points according to the data of 2020 increased by 2–8 mm/year (**Figure 3**). At the same time, transverse zoning is traced in the distribution of velocities, similar to the seismic one: to the west of Samur-Agdash velocity, disturbances are on average 8–10 mm/year, and to the east of it, they exceed 13 mm/year (13–15 mm/year).

At the same time, longitudinal zoning is observed in the distribution of the rates of horizontal movements, which correlates with the main Caucasian tectonic zoning of the territory.

Review of the distribution data of the velocity vectors of horizontal displacements of GPS geodetic points on the territory of Azerbaijan and the neighboring areas of Iran for the period 1998–2020 leads us to conclude about a significant (up to 15 mm/year) rate of movement in the north-north-east direction of the southwestern flank and the central strip of the South Caucasian microplate, including the territory of the south-eastern segment of the Lesser Caucasus, Kura depression, and Talysh. At the same time, within the northeastern flank of the microplate corresponding to the Vandam-Gobustan megazone of the Greater Caucasus, the velocity vectors are reduced to 6–13 mm/year, and even further north, in the hanging wing of the Kbaad-Zanginsky deep underthrust, that is, directly within the accretionary prism, completely decrease to 0–6 mm/year (data from 2010 to 2014). In general, the tangential contraction of the earth's crust in the region is estimated at 4–10 mm/year.

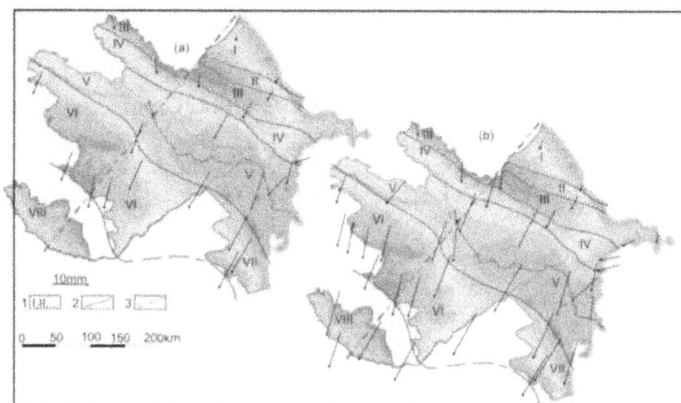

Figure 3.
GPS velocities of horizontal movements of the earth's surface in Azerbaijan and adjacent regions in 2004 (a) and 2020 (b). Compiled by R.T. Safarov. (1) main structural zones (longitudinal tectonic blocks): (I) Gusar-Davachinskaya, (II) the Lateral Ridge of the Greater Caucasus, (III) Southern slope of the Greater Caucasus, (IV) Kakheti-Vandam-Gobustan, (V) Kurinskaya, (VI) Artvin-Garabagh, (VII) Talysh, and (VIII) Araz; (2) deep faults at the boundaries of structural zones; and (3) Samur-Agdash fault.

This is confirmed by the observed directions and velocities of the earth's surface movement within territory of Azerbaijan and adjacent areas according to the results of measurements of GPS points in 2015–2020 (**Figure 4**). The velocity field clearly illustrates the movement of the earth's surface in the N-NE direction. At the same time, the plots clearly show a specific feature of the velocity field, namely a contrasting decrease in velocity at observation points located in the southern wing of the Zangin thrust fault, in comparison with the velocities recorded within the Kura and more southern zones (see **Figures 3** and **4**).

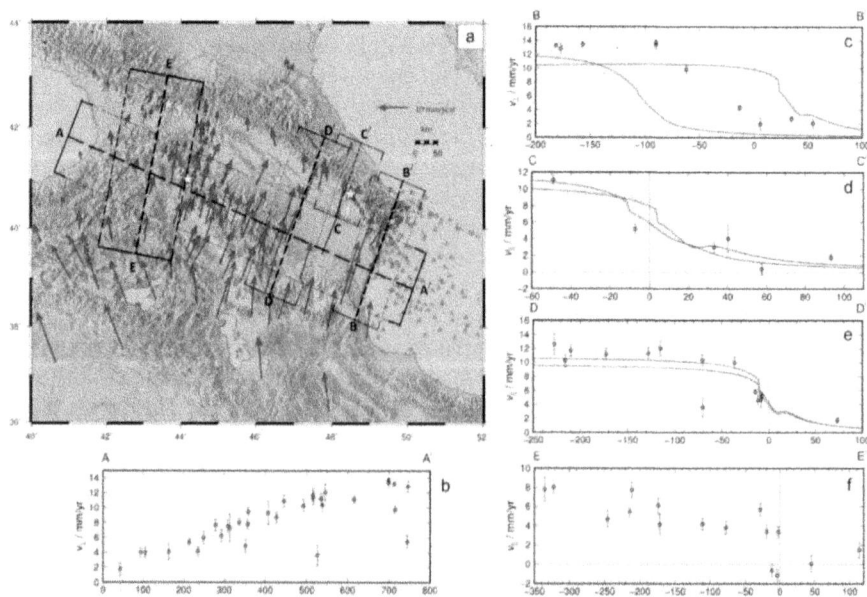

Figure 4.
GPS velocities of horizontal movements in Azerbaijan and adjacent regions (2020) and graphs of parallel and transverse components of GPS velocities along sections AA /, BB /, CC /, DD /, and EE / [5, 6, 16].

This phenomenon reflects the process of successive accumulation of elastic deformations in the pseudosubduction interaction zone of the northern flank structures of the South Caucasian microplate (Vandam-Gobustan megazone) with the accretionary prism of the Greater Caucasus.

Active faults: A well-pronounced indicator of the activity of faults (fault zones) is weak seismicity, so that, any even the smallest tectonic movements in disjunctive zones generate more or less strong seismic shakes.

The map of the earthquake epicenters that occurred on the territory of Azerbaijan over the past 20 years shows that focal zones are distributed very randomly here (**Figure 2**). At the same time, a similar peculiarity is observed within the most highly active regions, where the weaker earthquake foci clustering was observed in some areas.

One of the authors has developed a method for identifying real-time active segments of deep faults based on manifestations of weak seismicity in these zones [22]. This method is based on the idea of seismogenic structures (zones), which are known to be active faults that delimit geotectonic structures with different tectonic regimes and accumulate all strong and most of the weak and medium-strength earthquakes. According to the proposed method, the identification of seismogenic zones is carried out on the basis of the breakdown of the study area into equal areas and plotting of a map of weak seismicity. For each of these areas, within which the number of epicenters is not less than the specified threshold value, approximating lines of concentration of epicenters are constructed.

It is assumed that these lines correspond to active faults zones. These zones are actually potential source zones for strong earthquakes in specific territories. The method for determining active deep faults based on weak seismicity makes it possible to determine the location of potential source zones, as well as calculate their seismic potential and seismic effect that may occur on the earth's surface in the event of seismic activity. To assess the degree of their manifestation, the position of the sources of earthquakes and the parameters of the seismic regime are determined. At the same time, the catalogs of earthquakes are analyzed taking into account foreshock and aftershock activity, the stretch of pleistoseist zones, the character of the seismic effect decay depending on the distance, and other factors.

Coming from aforesaid, a map of potential seismic hazard for the territory of the Azerbaijan was compiled on the basis of a spatial analysis of weak seismicity (**Figure 5**), and active faults (fault zones) at the current stage of tectogenesis were identified. Based on the observed weak seismicity, active areas of these faults were identified, which are potential foci zones. At the same time, the relationship between the length of focal zones and the maximum possible magnitudes of earthquakes in them has been determined. It was found that the value of the maximum possible earthquake magnitudes (M_{MAX}) in the territory of Azerbaijan is approximately equal to 7 ($M_{MAX} = 6.9 \div 7.3$).

The features of the seismic activity manifestation on the territory of Azerbaijan are considered by us on the example of the southern slope of the Greater Caucasus, which at the current stage of tectogenesis is the most seismically active region of the country. Large seismic events periodically occur here, accompanied by the spontaneous release of large volumes of energy from the earth's interior. Seismic activity is associated with the ongoing intensive restructuring of the structural plan with significant amplitudes of the latest and modern movements: earthquake foci, as a rule, are confined to the boundaries of large geotectonic elements of the earth's crust and nodes of intersection of faults of various directions.

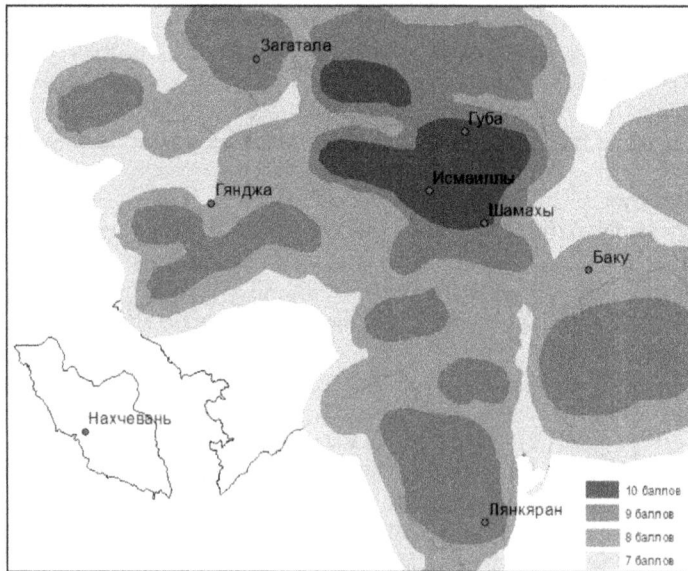

Figure 5.
Map of the potential seismic hazard of the territory of Azerbaijan. Compiled by T. Ya. Mammadli.

3. The dynamics of the manifestation of seismic activity

In-depth uneven distribution of earthquakes foci, in fact, proves ongoing pseudo-subduction interaction within southern slopes of the Greater Caucasus. The hypocentral levels exist in 2–6, 8–12, 17–22, and 25–45 km. The analysis of in-depth earthquake distribution evidences about existence of structural-dynamic interrelation between along with subvertical and subhorizontal contacts in the earth crust. Spatial and in-depth earthquake clustering can be explained from the point of view of block partibility and tectonic stratification of the earth crust (**Figures 6** and 7). Structurally, these clusters generally confine to the intersection junctions of fault zones with various directions or to the planes of tectonic ruptures and lateral displacements along weak contacts of multicomponent material complexes [21, 23–29].

Coming from temporal and spatial analysis of $M \geq 3$ earthquakes' foci distribution for the instrumental period of monitoring (1902–2020), we delineated dynamics of seismic activity in northern slope of the Greater Caucasus (**Figures 6** and 7). Using data of geophysical data reinterpretation, along with compiled tectonic and magmatic schemes of the study area, we divide this territory to four blocks (separated by various anti-Caucasian faults) with various levels of seismic activity [7, 29]. They are Zagatala, Sheki, Gabala-Shemakha, Gobustan zone. First, two clocks stand as eastern segments, whereas two others represent south-east segments of the Greater Caucasus. These segments are divided by Samur-Aghdash left-lateral strike-slip fault (**Figure 6**).

First, two blocks are distinguished for their lower seismic activity recorded throughout the entire period of observations (**Figure 7**):

- until 1980, 12 seismic events occurred within the Zagatala block's frontiers, confined to the consolidated crust's upper segment. Absolute majority of focuses (11) is located at depths of 12–30 km. Since 1980 until present, 66 events were recorded, with 9 events sourcing from the sedimentary cover, and 57—from 5 to 30 km depths of the consolidated crust;

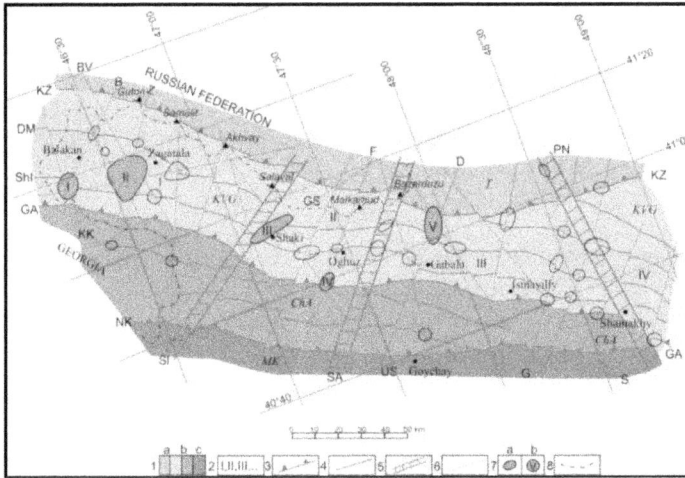

Figure 6.
Schematic map of fault tectonics and earthquakes foci zones distribution on the level of Pre-Jurassic basement—compiled by T.N. Kangarli, F.A. Aliyev and A.M. Aliyev [23]. (1) longitudinal blocks of the first order: (a) Tufan (T), (b) Kakheti-Vandam-Gobustan (KVG), (c) Chatma-Ajinohur (ChA), and (d) Middle Kur (MK); (2) transverse blocks of the first order: (I) Zagatala, (II) Shaki, (III) Gabala-Shamakhy, and (IV) Gobustan; (3) faults on borders of longitudinal blocks of the first range: KZ—Kbaad-Zangi, GA—Ganikh-Ayrichay-Alat, and NK—Northern Kur; (4) ruptures limiting the longitudinal blocks of the second order, including: DM—Dashaghil-Mudrisa and ShI—Shambul-Ismayilly; (5) faults on borders of transverse blocks of first level: Sl—Salavat, SA—Samur-Aghdash, and PN—Pirsaat-Neftchala; (6) other ruptures of anti-Caucasus direction, including: KK—Khimrikh-Khalatala, BV—Bulanligchay-Verkhiyan, B—Balakan, Z—Zagatala, GS—Gokhmug-Salyakhan, F—Fiy, US—Ujar-Saribash, D—Damiraparanchay, G—Girdimanchay, and S—Sighirly; and (7) earthquakes foci zones of 2012–2016 with M ≥ 3: (a) given in a paper: (I) Balakan, (II) Zagatala, (III) Shaki, (IV) Oghuz, and (V) Gabala; and (b) other; and (8) state border.

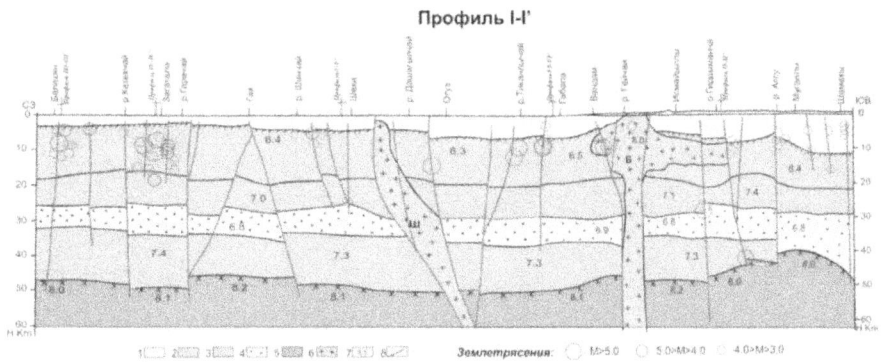

Figure 7.
Synthetic seismic profile of MOVZ along the longitudinal traverse of Balakian-Shamakhi. Compiled by T.N. Kangerli, A.M. Aliyev and F.A. Aliyev. 1–3 layers of the consolidated crust: (1) sedimentary; (2) "granite"; (3) "basalt"; (4) "waveguide"; (5) upper mantle; (6) intrusives (Ш—Sheki and B—Buinuz); (7) formation velocities of seismic waves; and (8) breaking violations.

First, two blocks differ by more pronounced seismic activity for the entire monitoring period (**Figure** 7). More detailed quantitative analysis gives us the following outcomes:

- Twelve seismic events took place within Zagatala block till 1980, which are confined to the upper part of earthquakes here. Nine of these events occurred in sedimentary layer, while 57 took place in consolidated crust within depth interval of 5–30 km;

- There occurred 14 seismic events within Shaki block till 1980. Three of these earthquakes took place in a depth of 3–5 km. which confines to alpine cover, while the rest part clustered in depths of 5–30 km, which confine with the consolidated crust. For the following period of 1981–2017, the number of earthquakes raised to 65, three of which occurred in the sedimentary layer, while 62 in consolidated crust (the distribution was 58 and 3 in the upper and lower segments respectively and 1 below Moho boundary).

Gabala-Shamakhy and Gobustan blocks have been more active throughout the entire period of observations, but there were also the leaps of seismic activity recorded in last quarter of XX century (**Figure 8**):

- until 1980, the total 29 seismic events have been registered within the block's structure, including the Alpine cover (14) and the consolidated crust's upper segment (14 events at depths of 5–30 km and 1 event below the Moho discontinuity). In the following period, the block's seismic activity increased to 219 events, 46 of which occurred in the sedimentary cover, 171—in the consolidated crust (141—5–30 km and 30—31–45 km), and 8—below the Moho discontinuity;

- Twenty-three seismic events have occurred in Gobustan block until 1980. Eight events were confined to the Alpine cover, 11—to the upper (5–30 km), and 4—to the lower segment (31–45 km) of the consolidated crust. During 1981–2017, the number of events increased to 196, 30 of which occurred in the sedimentary cover, 187—in the consolidated crust (139 in the upper and 48 in the lower segment), and 9—below the Moho discontinuity.

It can obviously be stated that the process of seismic activity was rising in the study area since 1980s of the last century. And this is despite the fact that the technical and methodological allowances for earthquake registration were not so qualitative as in the present (**Figure 9**). Within eastern segment of the study area, the upper part of the consolidated crust reveals as more seismic active, while in the south-eastern segment, earthquakes foci are scattered in the whole earth crust

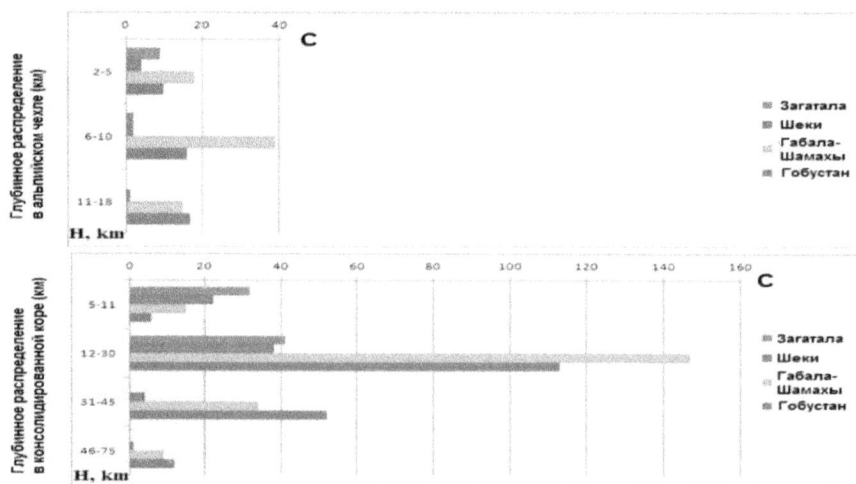

Figure 8.
Histogram of the vertical distribution of earthquake sources with M ≥ 3 over the blocks of the earth's crust on the southern slope of the Greater Caucasus within Azerbaijan (1902–2017). Compiled by F.A. Aliyev.

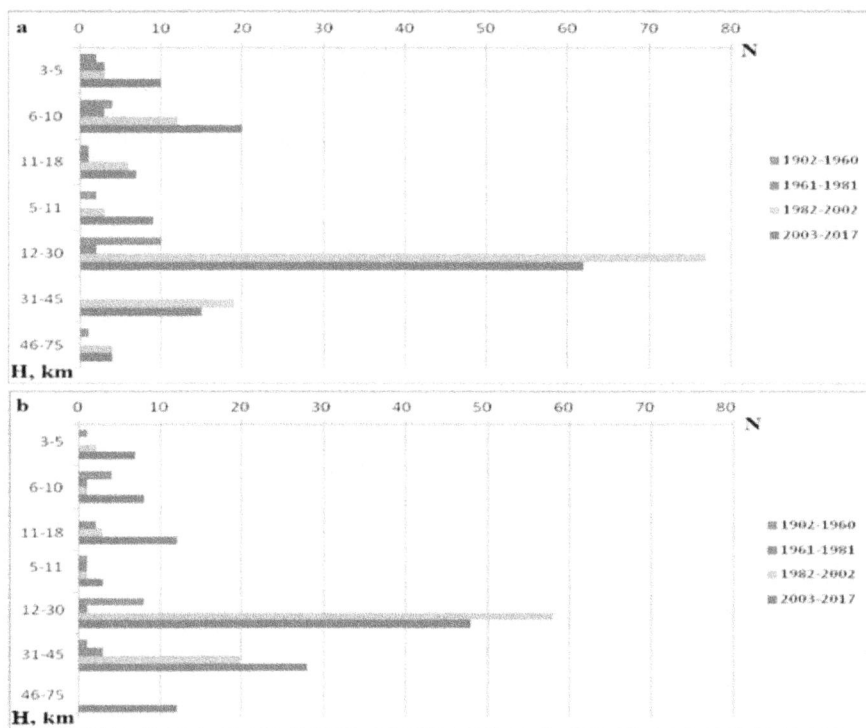

Figure 9.
Histogram reflecting changes in seismic activity (earthquakes with M ≥ 3) in the Azerbaijani part of the southern slope of the Greater Caucasus in space (in depth) and in time for the period 1902–2017: (a) Gabala-Shamakhi and Gobustan blocks; and (b) Zagatala and Sheki blocks. Compiled by F.A. Aliyev.

and also in the upper mantle. Most of deep seismic foci are located in a zone of the Western Caspian submeridianal fault. To the east of this fault zone, we can observe a stepped dipping the consolidated crust toward the Caspian hollow.

Spatial-temporal analysis of the distribution of strong seismic events in the Greater Caucasus region (within Azerbaijan borders) allows us to conclude that the northern flanks of Southern Caucasus microplate (these are structures that buried beneath accretionary wedge in the north, and the structures that revealed as a central segment or covered by a quaternary layer on the southern part of Kakheti-Vandam-Gobustan zone) are most active at the present stage of tectogenesis. These seismic active parts are divided into two zones:

- the zone controlled by the Ganykh-Airichay-Alat deep thrust of the all-Caucasian direction (or the border of Vandam-Gobustan and Sredne Kura tectonic zones) in the Ordubad-Balakyan-Mozdok strip of the anti-Caucasian seismotectonic zone in the west of the Azerbaijani part of the Greater Caucasus;

- complex tectonic node located in the east of described region within the boundaries of Talish-Samur-Makhachkala submeridional seismotectonic zone: corresponding to an intersection of two faulting zones: (1) northwest striking Western Caspian zone bordered by Pirsaat and Sighirly elementary deep strike slips from the northeast and southwest), and (2) northeast striking Girdimanchay-Shamakhy zone represented by Basgal-Khashi, Aghsu-Khaltan, and Jalair-Dibrar dislocations.

- a complex tectonic knot located within the Talysh-Samur-Makhachkala sub-meridional seismotectonic zone in the east of the described area, corresponding to the intersection of two fault zones: the northwestern direction West Caspian (bounded from the north-east by Pirsaat and south-west by Sygyrli by elementary right lateral strike-slip faults) and Girdymanchay-Gonagkend of north-eastern strike (represented by Basgal-Khashyn, Agsu-Khaltan, Sagiyan-Dibrar, Goylyardag-Nabur, and other disturbances).

Under lateral compression environment, small-scale blocks that constitute the region's earth crust trigger the emergence of transpressive deformations, which combine the shear displacements along framing transverse deformations with the compression structures such as "general-Caucasus" ruptures. Such regime leads to an emergence of multiple concentration areas of the elastic deformations confined to the mentioned dislocations and their articulation knots. It is just the exceeded ultimate strength of the rocks that causes an energy discharge and brittle destructions (according to stick-slip mechanism) in such tectonically weakened regions of the southern slope of the Azerbaijani part of Greater Caucasus (**Figure 6**).

Due to lateral compression state of the small-size blocks, into which the crust of the region is fragmented into parts, the formation of a transpressive type of deformation combines shear displacements along transverse faults. These faults confine blocks with compression structures, which include faults of the general Caucasian direction. In this tectonic regime, elastic stress is being concentrated in several zones that confined to the indicated dislocations and their junction points. Due to excess of the possible strength ability of rocks, the accumulated elastic deformations leads to energy discharge by means of earthquakes (mostly stick-slip type) in these tectonically weaker zones of the southern slope of the Greater Caucasus (see **Figure 10**).

The existence of tangential stresses in the region in real time is also indicated by the focal mechanisms of earthquakes with $M \geq 3$ that occurred in the period

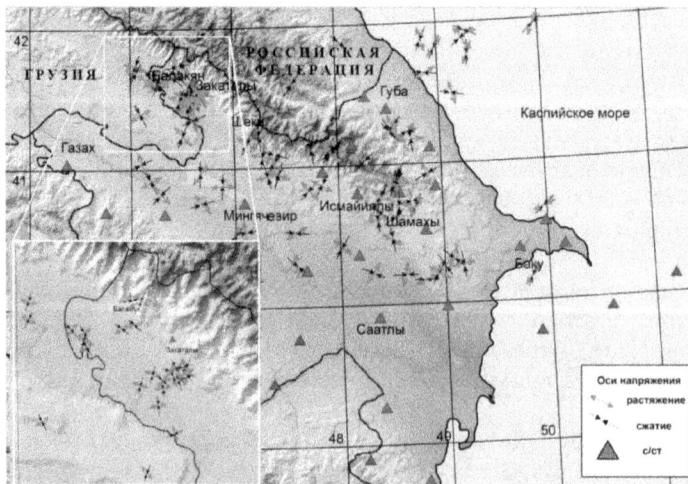

Figure 10.
Scheme of distribution of tectonic stresses from earthquake mechanisms with M ≥ 3 for 2003–2017—Compiled by S.E. Kazimova.

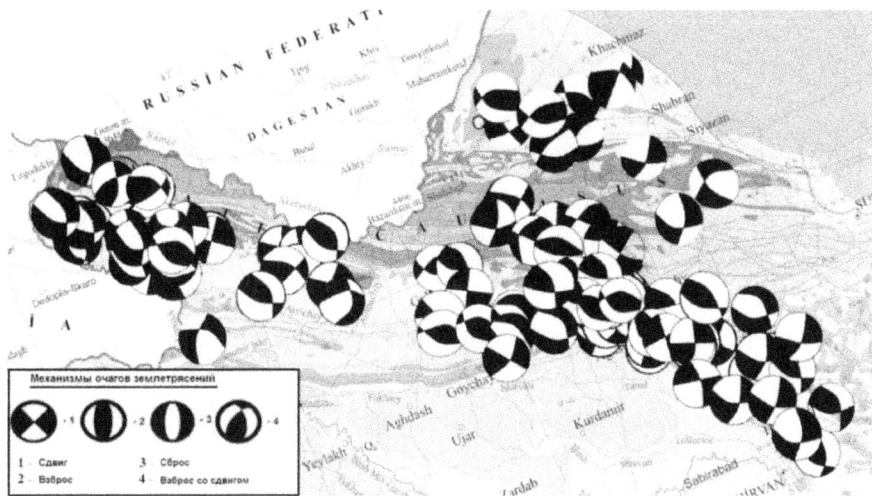

Figure 11.
Focal mechanisms of earthquakes with M ≥ 3 for the period 2003–2017. Compiled by S.E. Kazimova.

2003–2017 (**Figure 11**). Analysis of the distribution of the axes of compression and expansion indicates the predominance of lateral compression oriented in the submeridional and NE-SW directions. The types of focal mechanisms generally correspond to the concepts of the geodynamics of convergent microplate boundaries, where the entire set of these mechanisms is noted (**Figure 11**)—from fault to reverse [23–25, 27, 28, 30, 31].

This is evidenced by the seismic events that took place in Northwestern Azerbaijan in the period from 2012 to the first half of 2018. As an example, the conditions and factors that determined the tangible seismic activity in the Zagatala, Balakan, and Sheki source zones are given.

One of most seismically active zones in 2012–2015 was Zagatala focal zone, where three earthquakes with M = 5.27–5.69 (07.05.2012) took place, along with one event with M = 5.02 (20.06.2012) and numerous aftershocks with M = 3.0–4.4. A huge number of earthquakes foci are located within depth of 5–20 km, which confines to the pre-Jurassic basement of Kakheti-Vandam-Gobustan zone's frontal part. Despite the mostly clustered focal zone, one aftershock occurred outside of this zone, within Alpine cover (07.05.2012, 05:40). This is located in vicinity of Kvemo-Kedi village (Georgia) and corresponds to a plane of Ganikh-Ayrichay-Alat thrust fault that plunges in the northern rhumbs at its intersection with northeast-oriented Zagatala trans-tensional fault.

Overall, this source zone is a complex disjunctive node located in the upper part of the pre-Jurassic basement, consisting of elementary knots of intersection of tectonic faults with various orientations, where earthquake foci confined (see **Figures 12** and **13**). The volume of the rock mass, where earthquake hypocenters along with aftershocks with $M \geq 3$ are concentrated, reaches approximately 3400 km³.

These earthquake series are mainly associated with the activity of the Zagatala transverse fault, which in turn activated and related to the disturbances in the all-Caucasian and anti-Caucasian directions. Earthquake mechanisms here indicate the predominance of strike-slip and fault movements with the assistance of fault-strike-slip and reverse fault movements in the source zone.

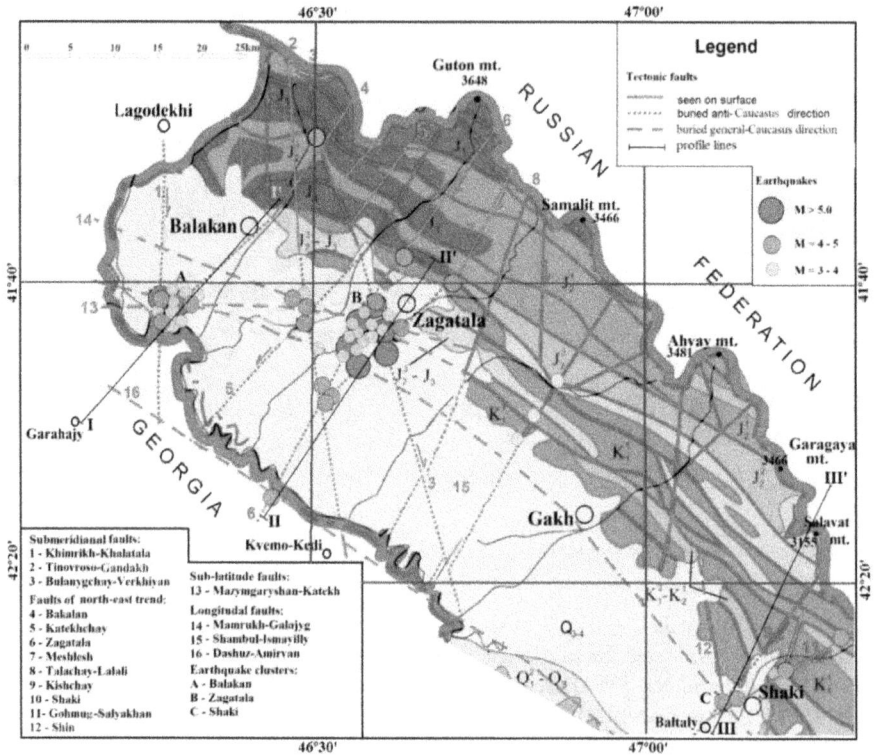

Figure 12.
The ratio of rupture dislocations and earthquake epicenters with M ≥ 3 for the period 2012–2014—By T.N. Kangarli, F.A. Aliyev, and A.M. Aliyev [23, 28].

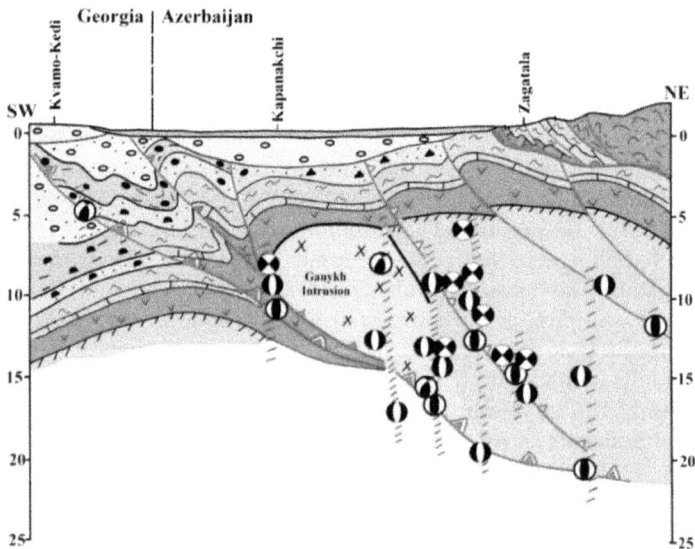

Figure 13.
Geological and geophysical section through Zagatala (III-III' in Figure 8, 14), Balakan focal zones of earthquakes—By T.N. Kangarli, F.A. Aliyev, and A.M. Aliyev [23, 28].

4. Conclusions

Analysis was performed and correlation between tectonics and modern seismic activity of the studied region leads to the following conclusions:

- under the influence of the introduction of the Arabian indenter into the buffer structures of the southern framing of Eurasia, the territory of the South Caucasus at the present stage of alpine tectogenesis (from the end of the Miocene) is an underthrust area (S-subduction—continental subduction) of the South Caucasian microplate under the southern underbelly of Eurasia (Turanian epigercyn platform) in the northern wing and active ground volcanism with the formation of volcano-plutonic complexes in the southern wing;

- analysis of the manifestation and distribution of weak seismicity allow to determine the position of active faults (fault zones) at the current stage and to calculate their seismic potential;

- recent seismic activization is generally confined with the data of horizontal displacement rates for the period of GPS observations of the modern geodynamics of the region; GPS monitoring data indicate an intensive advancement of the South Caucasian block in the northern points;

- analysis and correlation of historical and recent (2012–2020) seismic events indicate the confinement of earthquake foci mainly to the nodes of intersection of active ruptures of various strikes or to the planes of deep tectonic disruptions and lateral displacements along unstable contacts of material complexes of various competencies;

- focal mechanisms of seismic events reveal various ones, mainly near-vertical, planes of faults and strike-slip faults, but in general, the earthquake foci are confined to the nodes of intersection of faults of the general Caucasian and anti-Caucasian directions.

Author details

Talat Kangarli[1*], Tahir Mammadli[2], Fuad Aliyev[1], Rafig Safarov[1] and Sabina Kazimova[2]

1 Institute of Geology and Geophysics, ANAS, Baku, Azerbaijan

2 Republican Seismology Survey Center, ANAS, Baku, Azerbaijan

*Address all correspondence to: tkangarli@gmail.com

IntechOpen

References

[1] Dotduev SI. About napping structure of the Greater Caucasus. Geotectonics 1986;5:94-106 (in Russian) [Дотдуев С.И. О покровном строении Большого Кавказа // Геотектоника. 1986. №5. С. 94-106]

[2] Akhmedbeyli FS, Ismail-Zade AD, Kangarli TN. Geodynamics of the Eastern Caucasus in the alpine tectonic-magmatic cycle (Azerbaijan). In: Proceedings of the Institute of Geology of Azerbaijan National Academy of Sciences. Vol. 30. 2002. pp. 36-48 (in Russian) [Ахмедбейли Ф.С., Исмаил-заде А.Д., Кенгерли Т.Н. Геодинамика Восточного Кавказа в альпийском тектоно-магматическом цикле (Азербайджан) // Труды Института Геологии НАН Азербайджана. 2002. №30. С. 36-48]

[3] Allen M, Jackson J, Walker R. Late Cenozoic reorganization of the Arabia-Eurasia collision and the comparison of short-term and long-term deformation rates. Tectonics. 2004;23:TC2008. DOI: 10.1029/2003TC001530

[4] Geology of Azerbaijan. Tectonics. Vol. IV. In: Khain VE, Alizade A, editors. Baku: Nafta-Press Publishing House; 2005. 506 p (in Russian) [Геология Азербайджана, Т.IV. Тектоника (ред. В.Е.Хаина и Ак.А.Ализаде). Баку: Издательство «Nafta-Press», 2005. 506с.]

[5] Ismail-Zadeh A, Adamia S, et al. Geodynamics, seismicity, and seismic hazards of the Caucasus. Earth-Science Reviews. 2020;207:1-26. DOI: 10.1016/j.earscirev.2020.103222

[6] Kadirov FA, Mammadov SG, Safarov RT. Active geodynamics of the Caucasus. Geophysical Journal. 2017;39(4):98-101

[7] Kangarli TN. Mass overthrust within the structure of Greater Caucasus (Azerbaijan). In: The Modern Problems of Geology and Geophysics of Eastern Caucasus and the South Caspian Depression. Baku: Nafta-Press; 2012. pp. 163-201

[8] Khain VYe. Regional Geotectonics. Alpine Mediterranean Belt. Moscow: Publishing House Nedra; 1984. 344 p (in Russian) [Хаин В.Е. Региональная геотектоника. Альпийский Средиземноморский пояс. Москва: Издательство «Недра», 1984. 344с.]

[9] Khain VYe. Tectonics of Continents and Oceans (2000). Moscow: Publishing House "Scientific World"; 2001. 606 p (in Russian) [Хаин В.Е. Тектоника континентов и океанов (год 2000). Москва: Издательство «Научный мир», 2001. 606с.]

[10] Khain VYe, Grigoryans BV, Isayev BM. The West Caspian fault and some regularities of the manifestation of transverse faults in geosynclinal folded regions. Bulletin of the MIPT, Geol. Sect. 1966;2:5-23 [Хаин В.Е., Григорьянц Б.В., Исаев Б.М. Западно-Каспийский разлом и некоторые закономерности проявления поперечных разломов в геосинклинальных складчатых областях // Бюллетень МОИП, отд. геол., 1966, №2. С. 5-23]

[11] Khain VYe, Chekhovich PA. Main stages of tectonic development of the Caspian region. In: Khain VYe, Bogdanov NA, editors. International Tectonic Map of the Caspian Sea and its Surroundings. Scale 1:2500000. Explanatory Notes. Moscow: Publishing House "Scientific World"; 2006. pp. 57-64

[12] Kopp ML. Structures of Lateral Compression in the Alpine-Himalayan Collision Belt. Moscow: Publishing House "Scientific World"; 1997. 313 p (in Russian) [Копп М.Л. Структуры

латерального выжимания в Альпийско-Гималайском коллизионном поясе. Москва: Издательство «Научный мир», 1997. 313с.]

[13] Kopp ML. The modern structure of the Caspian region as a result of pressure of the Arabian plate. In: Geodynamics of the Black Sea-Caspian Sea Segment of the Alpine Fold Belt and Prospect of Mineral Exploration. Abstracts of the International Conference. Baku; 9-10 June 1999. Baku: Nafta-Press Publishing House; 1999. pp. 99-100 (in Russia) [Копп М.Л. Новейшая структура прикаспийского региона как результат давления Аравийской плиты // Геодинамика Черноморско-Каспийского сегмента Альпийского складчатого пояса и перспективы поисков полезных ископаемых. Тезисы докладов Международной конференции. Баку, 9-10 июня 1999г. Баку: Издательство «Nafta-Press», 1999. С. 99-100]

[14] Philip H, Cisternas A, Gvishiani A, Gorshkov A. The Caucasus: An actual example of the initial stages of continental collision. Tectonophysics. 1989;**161**:1-21. DOI: 10.1016/0040-1951(89)90297-7

[15] Vincent SJ, Morton AC, Carter A, Gibbs S, Teymuraz GB. Oligocene uplift of the western Greater Caucasus: An effect of initial Arabia-Eurasia collision. Terra Nova. 2007;**19**:160-166. DOI: 10.1111/j.1365-3121.2007.00731.x

[16] Kadirov FA, Floyd M, Reilinger R, Alizadeh AA, Guliyev IS, Mammadov SG, et al. Active geodynamics of the Caucasus region: Implications for earthquake hazards in Azerbaijan. Proceedings of the Azerbaijan National Academy of Sciences, Earth Sciences. 2015;**3**:3-17

[17] Kadirov FA, Kadyrov AG, Aliyev FA, Mamedov SK, Safarov RT. GPS-monitoring and seismicity of the collision zone of Azerbaijani part of the Greater Caucasus. Proceedings Azerbaijan National Academy of Sciences. The Sciences of Earth 2009;3:12-18 (in Russian) [Кадиров Ф.А., Кадыров А.Г., Алиев Ф.А., Мамедов С.К., Сафаров Р.Т. GPS-мониторинг и сейсмичность коллизионной зоны азербайджанской части Большого Кавказа // Известия НАН Азербайджана, Науки о Земле. 2009. №3. С. 12-18]

[18] Kadirov F, Mammadov S, Reilinger R, McClusky S. Some new data on modern tectonic deformation and active faulting in Azerbaijan (according to Global Positioning System Measurements). Proceedings of the Azerbaijan National Academy of Sciences, Earth Sciences. 2008;**1**:82-88

[19] Kadirov F, Safarov R, Mammadov S. Crustal deformation of the Caucasus region derived from GPS measurements. In: Proceedings of the 36th National and the 3rd International Geosciences Congress. 2018. Available from: http://36nigc.conference.gsi.ir/en

[20] Kadirov FA, Safarov RT. Deformation of the Earth's crust of Azerbaijan and adjacent territories based on GPS measurements. Proceeding Azerbaijan National Academy of Sciences. The Sciences of Earth 2013;1:47-55 (in Russian) [Кадиров Ф.А., Сафаров Р.Т. Деформация земной коры Азербайджана и сопредельных территорий по данным GPS-измерений // Известия НАН Азербайджана, Науки о Земле. 2013. №1. С.47-55]

[21] Telesca L, Kadirov F, Yetirmishli G, Safarov R, Babayev G, Ismaylova S. Statistical analysis of the 2003-2016 seismicity of Azerbaijan and surrounding areas. Journal of Seismology. 2017;**1467**:14-85. DOI: 10.1007/s10950-017-9677-x

[22] Mammadli TY. Identification of focal zones of strong earthquakes in

Azerbaijan and determination of their maximum magnitudes (M_{max}) by weak seismicity. ANAS Transactions, Earth Sciences. 2005;**1**:60-64

[23] Kangarli TN, Aliyev AM, Aliyev FA, Rahimov FM. A. Seismotectonic zoning of Azerbaijan territory. In: Proceeding of European Geosciences Union (EGU) General Assembly. 2017a. Available from: http://meetingorganizer. copernicus.org/EGU2017/EGU2017-12778.pdf

[24] Aliyev F, Kangarli T, Aliyev A, Vahabov U. Recent geodynamics and seismicity of the Greater Caucasus (within Azerbaijan borders). In: Proceedings of the 36th National and the 3rd International Geosciences Congress. 2018. Available from: http://36nigc.conference.gsi.ir/en

[25] Alizadeh AkA, Kangarli TN, Aliyev FA. Tectonic stratification and seismicity of the accretionary prism of the Azerbaijani part of Greater Caucasus. In: Proceeding of European Geosciences Union (EGU) General Assembly. 2013. Available from: http://meetingorganizer.copernicus.org/EGU2013/EGU2013-445-1.pdf

[26] Kangarli TN, Kadirov FA, Yetirmishli GJ, Aliyev FA, Kazimova SE, Aliyev AM, et al. Recent geodynamics, active faults and earthquake focal mechanisms of the zone of pseudosubduction interaction between the Northern and Southern Caucasus microplates in the southern slope of the Greater Caucasus (Azerbaijan). Geodynamics and Tectonophysics. 2018a;**9**(4):1099-1126. DOI: 10.5800/GT-2018-9-4-0385

[27] Kangarli TN, Aliyev FA, Rahimov FM, Murtuzov ZM. Tectonics, recent geodynamics and seismicity of Azerbaijan part of the Greater Caucasus. In: Proceeding of European Geosciences Union (EGU) General Assembly. 2016. Available from: http://meetingorganizer. copernicus.org/EGU2016/EGU2016-385-1.pdf

[28] Kangarli TN, Aliyev FA, Aliyev AM, Vahabov UG. B. Active tectonics and focal mechanisms of earthquakes in the pseudosubduction active zone of the North- and South-Caucasus microplates (within Azerbaijan). Geophysical Journal. 2017b;**39**(4):101-104

[29] Kangarli TN, Veliev GO. Direction and results of the Ismail-Shemakha study area in relation with seismic prediction. In: Forecast of Earthquakes. Vol. 10. Dushanbe, Moscow: Publishing House "Donish"; 1988. pp. 172-185 (in Russian) [Кенгерли Т.Н., Велиев Г.О. Направление и результаты исследований Исмаилы-Шемахинского полигона в связи с сейсмопрогнозом // Прогноз землетрясений. Душанбе, Москва: Издательство «ДОНИШ», 1988. №10. C. 172-185]

[30] Rzayev AG, Yetirmishli GD, Kazymova SE. Reflection of the geodynamic regime in variations of the geomagnetic field intensity (on example of the southern slope of the Greater Caucasus). Proceeding Azerbaijan National Academy of Sciences. The Sciences of Earth 2013;4:3-15 (in Russian) [Рзаев А.Г., Етирмишли Г.Д. Казымова С.Э. Отражение геодинамического режима в вариациях напряженности геомагнитного поля (на примере южного склона Большого Кавказа) // Известия НАН Азербайджана, Науки о Земле. 2013. №4. C. 3-15]

[31] Yetirmishli GJ, Kazimova SE, Ismailova SS, Garaveliyev ES. Dynamic and kinematic characteristics of earthquakes of Sheki-Oguz region. Proceedings of the Azerbaijan National Academy of Sciences, Earth Sciences. 2016;**3-4**:28-36